797,885 Books
are available to read at

www.ForgottenBooks.com

Forgotten Books' App
Available for mobile, tablet & eReader

ISBN 978-1-5277-4169-0
PIBN 10886307

This book is a reproduction of an important historical work. Forgotten Books uses state-of-the-art technology to digitally reconstruct the work, preserving the original format whilst repairing imperfections present in the aged copy. In rare cases, an imperfection in the original, such as a blemish or missing page, may be replicated in our edition. We do, however, repair the vast majority of imperfections successfully; any imperfections that remain are intentionally left to preserve the state of such historical works.

Forgotten Books is a registered trademark of FB &c Ltd.
Copyright © 2017 FB &c Ltd.
FB &c Ltd, Dalton House, 60 Windsor Avenue, London, SW19 2RR.
Company number 08720141. Registered in England and Wales.

For support please visit www.forgottenbooks.com

1 MONTH OF FREE READING

at

www.ForgottenBooks.com

By purchasing this book you are eligible for one month membership to ForgottenBooks.com, giving you unlimited access to our entire collection of over 700,000 titles via our web site and mobile apps.

To claim your free month visit: www.forgottenbooks.com/free886307

* Offer is valid for 45 days from date of purchase. Terms and conditions apply.

English
Français
Deutsche
Italiano
Español
Português

www.forgottenbooks.com

Mythology Photography **Fiction**
Fishing Christianity **Art** Cooking
Essays **Buddhism** Freemasonry
Medicine **Biology** Music **Ancient Egypt** Evolution Carpentry Physics
Dance Geology **Mathematics** Fitness
Shakespeare **Folklore** Yoga Marketing
Confidence Immortality Biographies
Poetry **Psychology** Witchcraft
Electronics Chemistry History **Law**
Accounting **Philosophy** Anthropology
Alchemy Drama Quantum Mechanics
Atheism Sexual Health **Ancient History**
Entrepreneurship Languages Sport
Paleontology Needlework Islam
Metaphysics Investment Archaeology
Parenting Statistics Criminology
Motivational

TM 9-227

WAR DEPARTMENT

TECHNICAL MANUAL

20-MM AUTOMATIC GUN M1
AND
20-MM AIRCRAFT AUTOMATIC GUN AN-M2

NOVEMBER 19, 1942

*TM 9-227

TECHNICAL MANUAL }   WAR DEPARTMENT,
No. 9-227   Washington, November 19, 1942.

20-MM AUTOMATIC GUN M1
AND
20-MM AIRCRAFT AUTOMATIC GUN AN-M2

Prepared under the direction of the
Chief of Ordnance

CONTENTS

			Paragraphs	Pages
SECTION	I.	Introduction	1-5	3-7
	II.	Description and functioning	6-20	8-34
	III.	Operation	21-27	35-41
	IV.	Malfunctions and immediate action	28-29	42-43
	V.	Care and preservation	30-34	44-45
	VI.	Disassembly and assembly	35-44	46-73
	VII.	Inspection	45-57	74-78
	VIII.	Ammunition	58-72	79-90
	IX.	Organization spare parts and accessories	73-74	91-92
	X.	Storage and shipment	75-77	93
	XI.	Operation under unusual conditions	78-80	94-95
	XII.	References	81-82	96
INDEX				97-101

* This manual supersedes the following publications: TM 9-227, April 2, 1942, TB 227-1, January 1, 1942, TB 227-2, February 6, 1942 and TB 227-3, March 21, 1942.

TM 9-227

**20-MM AUTOMATIC GUN M1 AND
20-MM AIRCRAFT AUTOMATIC GUN AN-M2**

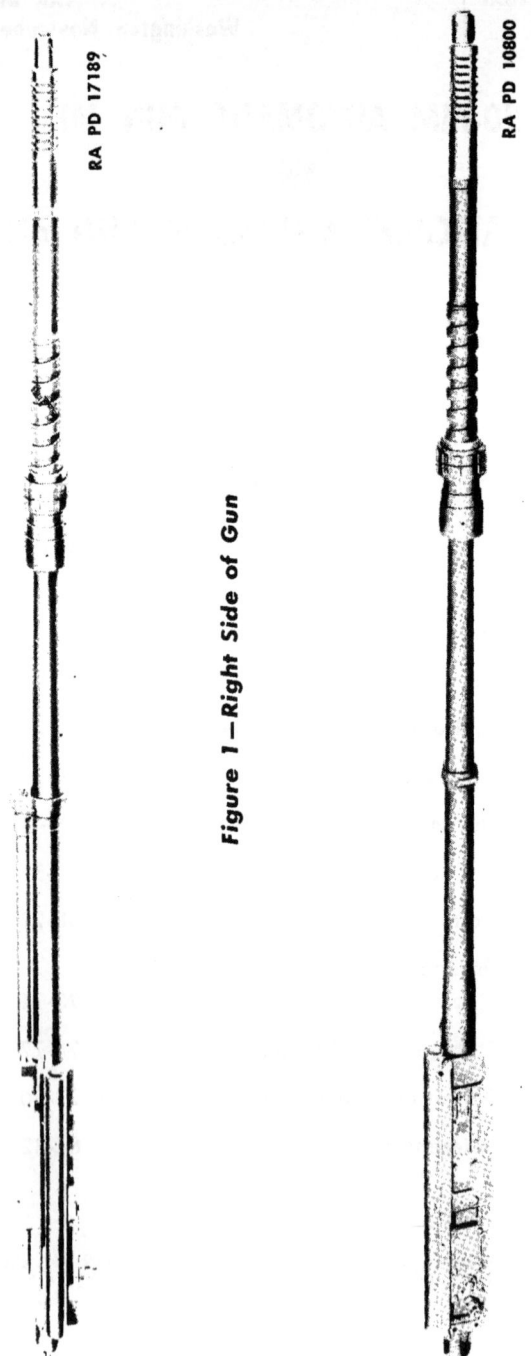

Figure 1 — Right Side of Gun

Figure 2 — Bottom of Gun

2

TM 9-227
1-3

Section I

INTRODUCTION

	Paragraph
Scope	1
Characteristics	2
Differences among models	3
Data	4
Cautions	5

1. SCOPE.

   a. This manual is published for the information and guidance of the using arms and services. It contains information of a technical nature required by the using arms for the identification, use, care, and preservation of the 20-mm Automatic Gun M1 and 20-mm Aircraft Automatic Gun AN-M2, and of the accessories and ammunition used therewith.

   b. Information on charges and solenoids will be published when available.

   c. This manual differs from previous issues of TM 9-227, 20-mm Aircraft Gun Materiel M1 and M2 as follows:

   (1) The table of data has been enlarged.
   (2) A list of cautions has been inserted in section I.
   (3) Treatment of functioning is much more detailed.
   (4) Instructions on lubrication of ammunition are included in section III as well as the loading of both the 60-round magazine and the Feed Mechanism M1.
   (5) Section IV, Malfunctions and Immediate Action, has been added.
   (6) Section VII, Inspection, has been added.
   (7) Section X, Storage and Shipment, has been added.
   (8) Section XI, Operation Under Unusual Conditions, has been added.

2. CHARACTERISTICS.

   a. The 20-mm Automatic Gun is a combination blowback and gas-operated aircraft weapon. The gun is air-cooled and has a cyclic rate of fire of 600 to 700 rounds per minute. It is designed for mounting as a fixed gun in the wing or fuselage of an airplane. It may also be mounted to fire through the hub of the propeller (figs. 1, 2, 3, and 4).

3. DIFFERENCES AMONG MODELS.

   a. The differences between the AN-M2 and M1 guns are in manufacture only; these do not affect troop use or care, but are useful as means of identifying the different models. The guns are identical with

20-MM AUTOMATIC GUN M1 AND
20-MM AIRCRAFT AUTOMATIC GUN AN-M2

respect to the construction of the tube and the working parts, the only differences being in the dimensions of some of the receiver parts. The AN-M2 Receiver is 0.2 inches longer. Each receiver slide of the AN-M2 Gun has a projection which fits into a slot in the receiver side, and the receiver slide bolts are locked by cotter pins. In the M1 Gun, each receiver slide has a head flange which overlaps the bottom face of the receiver side, and the receiver slide bolts are locked by locking wire. In some M1 Guns, the receiver slides have no head flanges and are riveted instead of bolted to the receiver. The shoulders on the bottom faces of the receiver slides serve as further means of identifying the M1 Gun.

NOTE: The AN-M2 Gun is the new designation for the M2 Gun with which the using arm is already provided. The M2 and the AN-M2 Guns are identical.

4. DATA.

Weight of M1 Gun without magazine	112 lb
Weight of AN-M2 Gun without magazine	118 lb
Weight of tube	47.5 lb
Weight of 60-Round 20-mm M1 Magazine	22 lb
Weight of 20-mm M1 Feed Mechanism	19 lb
Over-all length of gun	100.6 in.
Length of tube	67.5 in.
Muzzle velocity	2850 fps

Rifling:

Number of grooves	9
Depth of grooves	0.015 in.
Width of grooves	0.205 in.
Width of lands	0.068 in.
Twist, uniform, right-hand, slope	7 deg
Length	63.08 in.

Bore of Tube:

Across rifling lands	0.787 in.
Across rifling grooves	0.817 in.
Powder pressure (maximum)	42,000 psi
Travel of projectile in tube	63.68 in.

Rate of fire	600-700 rounds per min
Maximum allowable recoil	1.181 in.
Minimum recoil to operate M1 Feed Mechanism	0.787 in.
Ideal recoil to operate M1 Feed Mechanism	0.945 in.

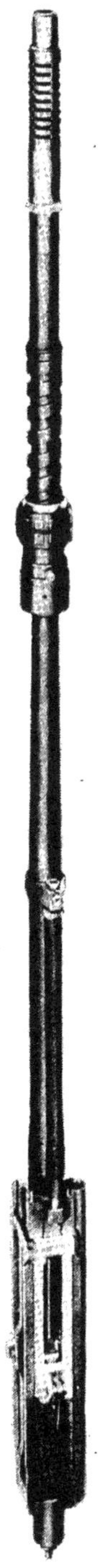

INTRODUCTION

RA PD 17191

RA PD 17190

Figure 3—Top of Gun

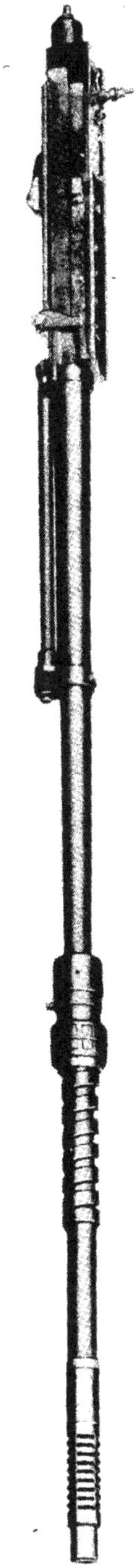

Figure 4—Left Side of Gun

20-MM AUTOMATIC GUN M1 AND
20-MM AIRCRAFT AUTOMATIC GUN AN-M2

5. CAUTIONS.

a. The using service will find that in a great many cases certain parts of the gun will be replaced, removed, or modified to facilitate installation. When this occurs, do not attempt to replace a strange part or to modify the gun assembly according to the instructions in this training manual. The modification or replacement of certain minor exterior gun parts is done to accommodate special airplanes and should not cause confusion.

h. All rounds should be lubricated just before they are inserted in the magazine or belt. Dip cloth in OIL, lubricating, preservative, light, and wipe the cartridge case with it, depositing a light film of oil evenly over the case and taking care not to oil the base of the cartridge case or the projectile. If OIL, lubricating, preservative, light, is not available, use OIL, lubricating, for aircraft instruments and machine guns.

c. Place the breechblock in the most forward (locked) position whenever the gun is to be disassembled. This is to reduce the tension on the driving spring and prevent possible injury when removing the spring.

d. Make certain that the gun is cocked before take-off.

e. Do not try to remedy a stoppage by recocking the gun in flight and attempting to fire. This may cause a high explosive round to strike the base of an unextracted round and result in an explosion.

INTRODUCTION

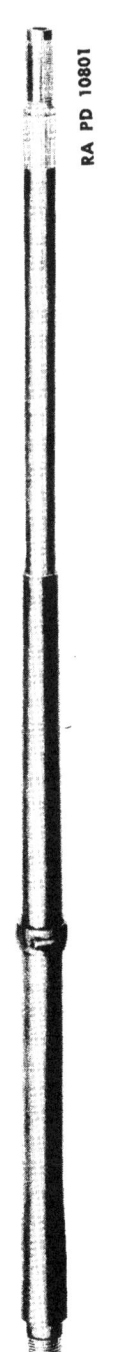

Figure 5 — Tube of Gun

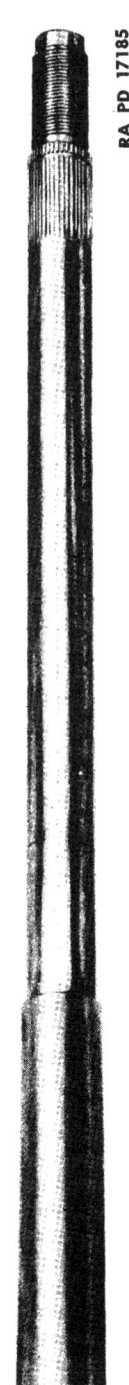

Figure 6 — Front Portion of Tube

TM 9-227
6–7

20-MM AUTOMATIC GUN M1 AND 20-MM AIRCRAFT AUTOMATIC GUN AN-M2

Section II

DESCRIPTION AND FUNCTIONING

	Paragraph
Tube	6
Muzzle brake assembly	7
Recoil spring and mounting sleeve group	8
Gas cylinder and sleeve group	9
Receiver assembly	10
Magazine slide group	11
Breechblock assembly	12
Breechblock locking key	13
Sear block group	14
Sear cover plate group	15
Rear buffer assembly	16
Driving spring guide group	17
20-mm Feed Mechanism M1	18
20-mm 60-Round Magazine M1	19
Functioning of the gun as a whole	20

6. TUBE.

a. The principal components of the 20-mm Automatic **Gun M1** and 20-mm Aircraft Automatic **Gun AN-M2** are the tube which accommodates the recoil mechanism and the receiver which houses most of the working parts. The tube is threaded at both ends and has a radial gas port drilled about 20 inches from the breech end. The breech end is screwed into the receiver and is secured with a locking pin to prevent its vibrating loose during firing. The breech face of the tube is recessed to clear the lip of the extractor (figs. 5 and 6).

7. MUZZLE BRAKE ASSEMBLY.

a. The muzzle end of the tube accommodates the muzzle brake assembly which counteracts some of the recoil. The muzzle brake assembly consists of a front ferrule, body assembly locking washer, and rear ferrule. The body assembly is composed of a sleeve and 8 baffles staked in place. It has 36 equally spaced ports cut at an angle of 45 degrees to the axis of the bore. This construction causes a portion of the blast gases to be deflected to the rear, thus absorbing about 35 percent of the recoil action (fig. 7).

b. The muzzle brake is used only with the 60-round magazine. When the feed mechanism is used the muzzle brake is removed and replaced

DESCRIPTION AND FUNCTIONING

with the thread protector ($_{fig.}$ 8) as a recoil of 0.787 plus inch is needed to operate the mechanism.

c. Just to the rear of the muzzle threads are a number of longitudinal splines which provide a locking surface for the muzzle brake lock. The muzzle brake lock and rear ferrule of the muzzle brake interlock by means of dentals on the lock and ferrule.

8. RECOIL SPRING AND MOUNTING SLEEVE GROUP.

a. Between the muzzle brake lock and the tube shoulder, the tube mounts the recoil spring and mounting sleeve group. The mounting sleeve assembly serves two purposes: to mount the front of the gun in the airplane and to act as a buffer during counterrecoil of the gun. It consists of the mounting sleeve nut slipped over the mounting body which, in turn, is screwed into the dash pot cylinder containing the piston and fiber washer. The mounting sleeve nut is threaded on the inside to secure it to the stationary mount in the airplane. The detent around the nut acts as a safety lock and prevents the nut from vibrating loose from the mount during firing (figs. 7 and 8).

b. The bushing, located between the mounting body and the dash pot cylinder, provides a bearing for the assembly on the front of the piston. The bushing is oiled through the oil plug which also acts to prevent the mounting body and dash pot cylinder from unscrewing. The short end of the dash pot piston rides against the shoulder on the tube while the long end rides against the rear end of the recoil spring filler sleeve ($_{fig.}$ 54). The front end of this sleeve rests against the flange of the recoil spring sleeve. The function of the filler sleeve is to form a lining for the recoil spring and to insure that the dash pot piston recoils with the gun. The recoil spring bears between the shoulder in the mounting body and the flange on the recoil spring sleeve which, in turn, abuts the muzzle brake lock. Initial compression is applied to the recoil spring when the muzzle brake or thread protector is screwed on in position. The recoil spring acts also as a recuperating spring.

c. As the gun recoils, the piston is drawn to the rear of the cylinder and creates a vacuum in front of the piston and thus lessens the recoil. As the gun recoils backward far enough, the piston passes the eight ports in the cylinder and air is taken into the cylinder. When the gun moves forward, the piston traps and compresses the intaken air which buffers the shock of counterrecoil.

d. A new recoil mechanism, designated 20-mm Gun Adapter AN-M1, will replace the recoil mechanism in many installations. The new

TM 9-227

20-MM AUTOMATIC GUN M1 AND 20-MM AIRCRAFT AUTOMATIC GUN AN-M2

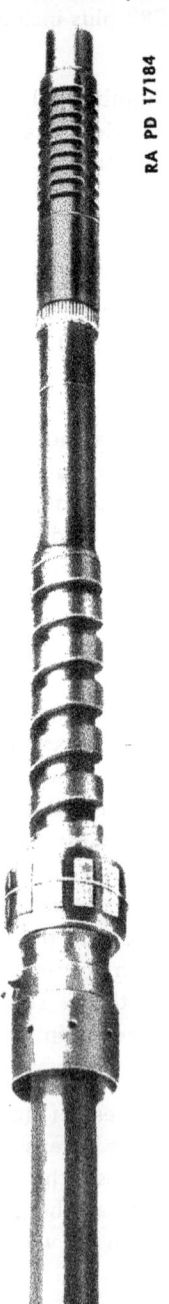

Figure 7 — Muzzle Brake, Recoil Spring and Front Mounting Group

Figure 8 — Thread Protector, Recoil Spring and Mounting Group

DESCRIPTION AND FUNCTIONING

adapter is a self-contained cylinder which slips over the barrel of the gun and absorbs recoil energy by reason of its integral ring spring. The adapter comprises a recoil spring together with a ring spring recoil and counterrecoil buffer, all enclosed in a metal cartridge surrounding the tube. A front spacer sleeve, a tube locking nut, and a front spacer shell nut locking spring are required to secure the adapter to the gun. Since both types of recoil mechanisms will be found in service, it is necessary that all personnel be familiar with both types. Information regarding the adapter will be published when available.

9. GAS CYLINDER AND SLEEVE GROUP.

a. The gas cylinder and sleeve group are shown assembled to the gun in figure 10. This group consists of a piston integral with a sleeve extension terminating in the form of a yoke which engages two push rods projecting through the front of the receiver, a gas cylinder guide which supports the sleeve, a cylinder which contains and guides the piston, a piston return spring contained in the sleeve, and bracket and vent plugs which close the openings of the gas cylinder bracket and secure the gas cylinder to the bracket. These parts are shown disassembled in figure 56.

10. RECEIVER ASSEMBLY.

a. The receiver assembly consists of the receiver body and the receiver plate which is riveted to the rear underside (figs. 10 and 11). The receiver houses most of the working parts. It also supports the feeder and serves for attaching the gun to the rear mounting. At the front end, the body is threaded internally to receive the tube; a vertical hole is drilled from the underside to accommodate the tube locking pin. On top of the body is a lug which is threaded internally to receive the gas cylinder guide. On each side below the lug, longitudinal holes are drilled through the front of the body to house the push rods which unlock the breechblock lock. Integral with the right side of the receiver is a charging cylinder which can be fitted with a manual or hydraulic charging unit for retracting the breechblock. A slot in the rear half of the charging cylinder enables the flange on the charging piston or the manual charging unit to engage the lug on the right breechblock slide. The front underside of the body is open to permit empty cartridge cases to be ejected. Above the ejector opening are two receiver slides, which are bolted or riveted to the sides and serve to support the breechblock in its forward movement. The slides have cammed surfaces at the rear which engage corresponding cams on the breechblock lock so as to lower it to the locked position. To the rear of the ejector opening a slot is cut in each side of the receiver body to accommodate the breechblock lock-

TM 9-227
10

20-MM AUTOMATIC GUN M1 AND
20-MM AIRCRAFT AUTOMATIC GUN AN-M2

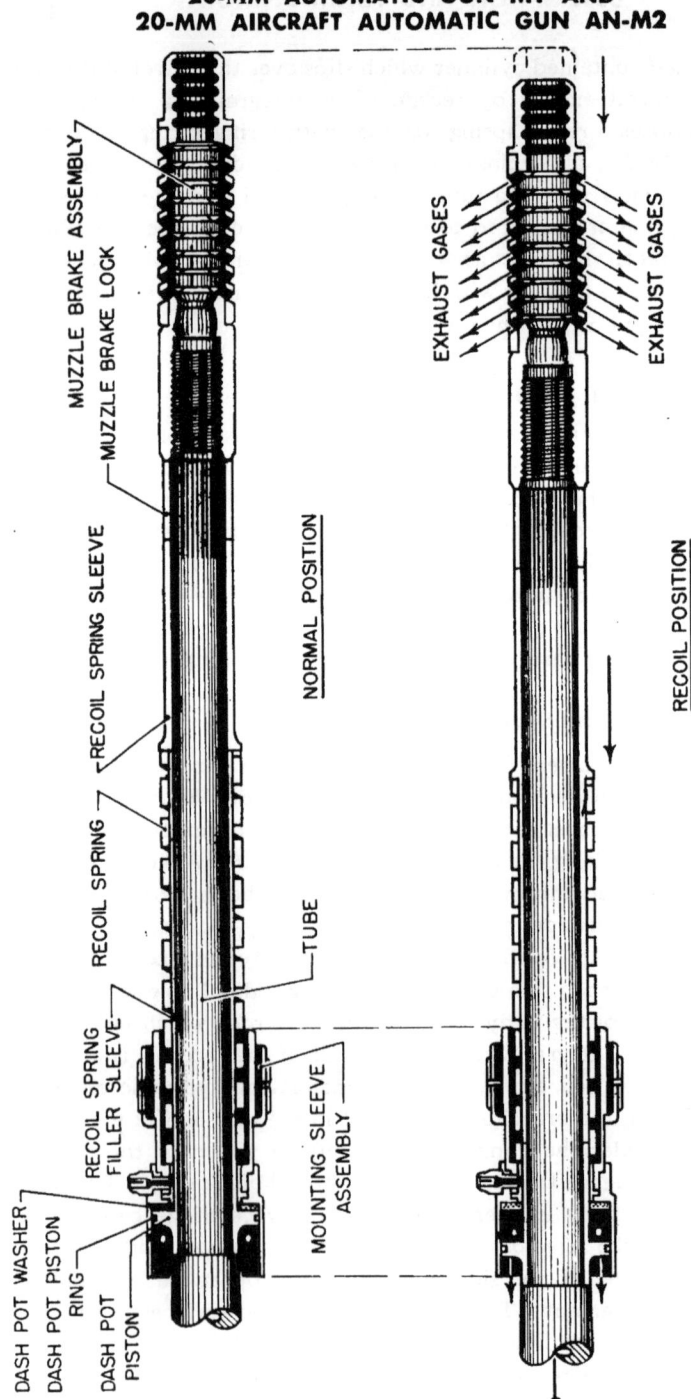

Figure 9 — Action of Recoil Mechanism

12

DESCRIPTION AND FUNCTIONING

ing key which allows the breechblock lock to hinge down and engage in front of it. At the rear, the underside of the receiver is partially closed by the receiver plate. The plate is shaped to house the sear mechanism and to accommodate the sear cover plate assembly with the firing parts. The rear of the body has vertical dovetail grooves for attaching the rear buffer assembly. The top of the receiver body has an opening with guideways for attaching the magazine slide group. The magazine slide group secures the magazine and also mounts the ejector.

11. MAGAZINE SLIDE GROUP.

a. The magazine slide has a guide on each side which provides for sliding engagement with corresponding guideways on the receiver body (figs. 10 and 55). The left front side of the slide supports a securing arm, the purpose of which is to anchor the slide rigidly so that it will remain stationary as the gun recoils. In a typical installation, the securing arm is bolted to a stationary part of the mount or airplane structure. In other installations, however, the securing arm has been removed and replaced by an adjustable tie bar bolted directly to the slide and connected with the airplane structure.

h. At the front of the slide there are two hook-shaped projections which secure the front of the magazine. Two longitudinal grooves at the rear of the slide accommodate the ejector and magazine latch.

c. The magazine latch fits in the grooves above the ejector (fig. 55). The latch houses two springs which abut the magazine slide backplate and keep the latch under tension. The latch is operated by the magazine slide lever which is supported on two ears on the magazine slide by the magazine slide lever pin and bushing.

d. The ejector also fits into the grooves in the magazine slide beneath the latch. It consists of two horns integral with a steel plate. The ejector plate houses two springs which contact the magazine slide backplate and keep the ejector forward. It is fitted with a threaded stud which passes through the backplate and is secured to it by a nut.

e. The upper inner surfaces of the horns are shaped to center and support the incoming round in the path of the breechblock as it moves forward. The forward ends of the two horns position the top of the bolt and deflect the empty cartridge case downward.

12. BREECHBLOCK ASSEMBLY.

a. The breechblock assembly consists of the bolt assembly, two breechblock slides with inertia blocks and suitable plungers, plunger

TM 9-227
12

**20-MM AUTOMATIC GUN M1 AND
20-MM AIRCRAFT AUTOMATIC GUN AN-M2**

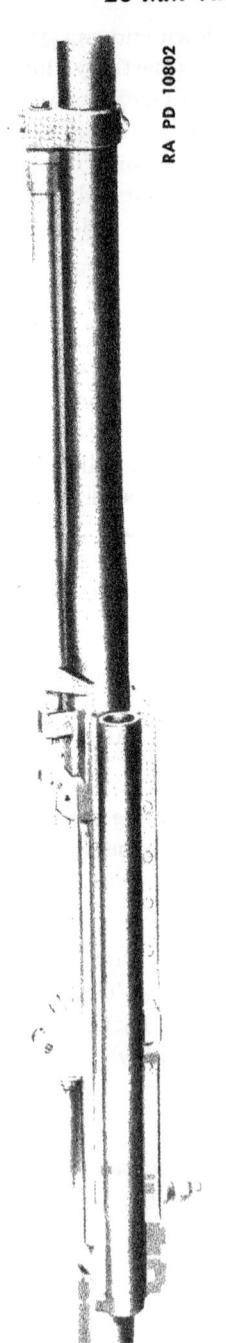

Figure 10 — Receiver and Rear Portion of Tube

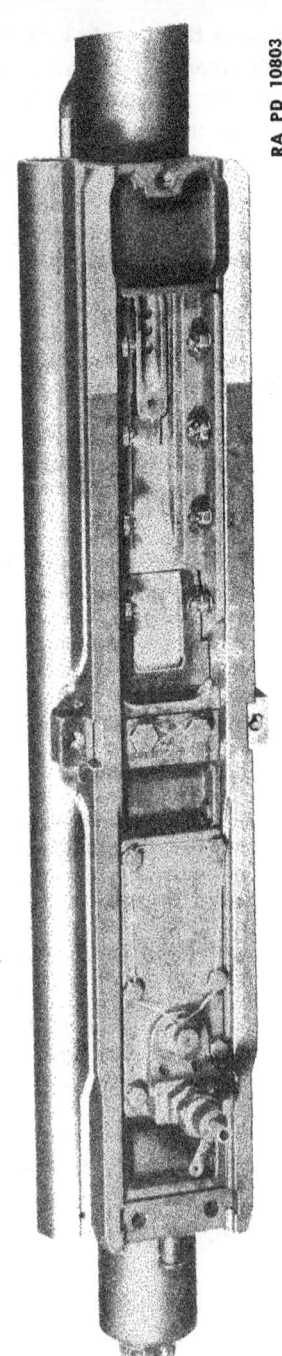

Figure 11 — Bottom of Receiver

DESCRIPTION AND FUNCTIONING

springs and guides, firing pin, breechblock lock, and extractor assembly. The whole group is housed in the receiver; its function is to carry the round from the mouth of the magazine into the chamber and to fire the round while holding it in the chamber (figs. 12, 13, 14 and 15).

b. The bolt is bored from the rear to receive the firing pin, driving spring guide plunger, and driving spring. The upper edges of the bolt are cut away to allow it to pass between the horns of the ejector, while the front face is recessed to accommodate the base of the cartridge case. The bottom of the bolt is recessed at the rear to receive the breechblock lock and at the front for securing the extractor. The extractor is attached by a pivot pin. A spring between the extractor and the bolt forces the claw at the forward edge of the extractor toward the face of the bolt.

c. Flanges along the lower edges of the bolt provide surfaces for guiding the breechblock slides. The slides are keyed together by means of a slide key passing through a slot near the forward end of the bolt. The breechblock slide key mates with a transverse slot in the firing pin so that the slides and pin move together as a unit. The lug on the rear end of the right breechblock slide extends through a slot into the cylinder for engagement with the charging unit. Each breechblock slide spring guide bears between the breechblock pin and a recess in the slide. The spring which is mounted around the guide helps to drive the slides forward in the firing position and to prevent rebound of the slides. The bottom edges near the rear of the slides are cut to form cam surfaces which contact corresponding surfaces on the breechblock lock.

d. A large slot in each breechblock slide accommodates an inertia block. The inertia blocks are cut away on the underside to accommodate the breechblock slide springs and guides and are drilled at the front to house a plunger and spring. The plunger bears against the slot in the slide. A recess in the plunger accommodates the pin which holds it in position and limits its movement. The function of the inertia blocks is to prevent rebound of the breechblock slides. The shallow grooves on the slides and inertia blocks distribute the lubricant and collect any foreign matter.

e. The breechblock lock (fig. 47) is a flat plate with lugs projecting from each side of its top surface. The rear surfaces of the lugs are also cammed to contact the beveled surface of the breechblock locking key. The half-cylindrical front edge hinges in the recess on the underside of the bolt. The bottom of the lock is recessed for engaging the sear.

13. BREECHBLOCK LOCKING KEY.

a. The breechblock locking key is housed in the transverse slot in the receiver (fig. 57). It is prevented from moving sideways by the plate

TM 9-227
13

**20-MM AUTOMATIC GUN M1 AND
20-MM AIRCRAFT AUTOMATIC GUN AN-M2**

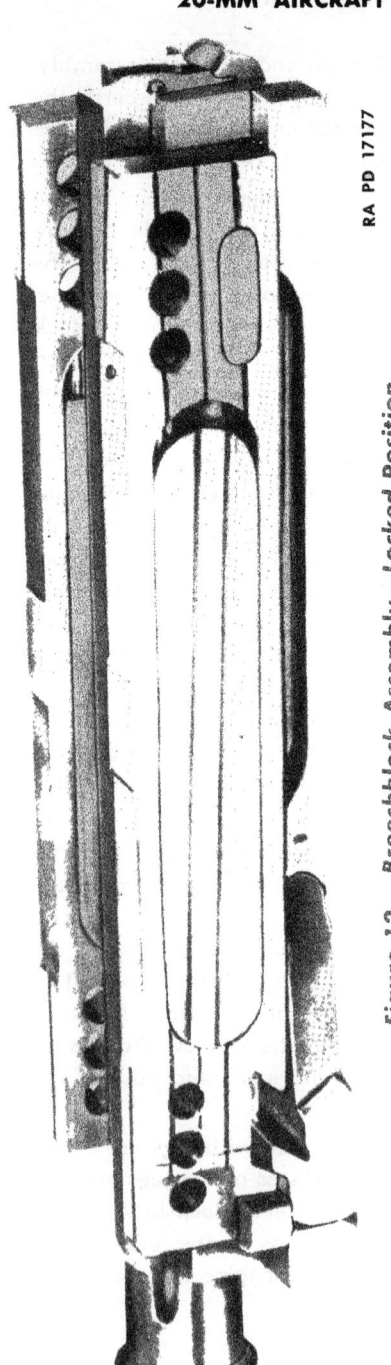

Figure 12 — Breechblock Assembly — Locked Position

RA PD 17177

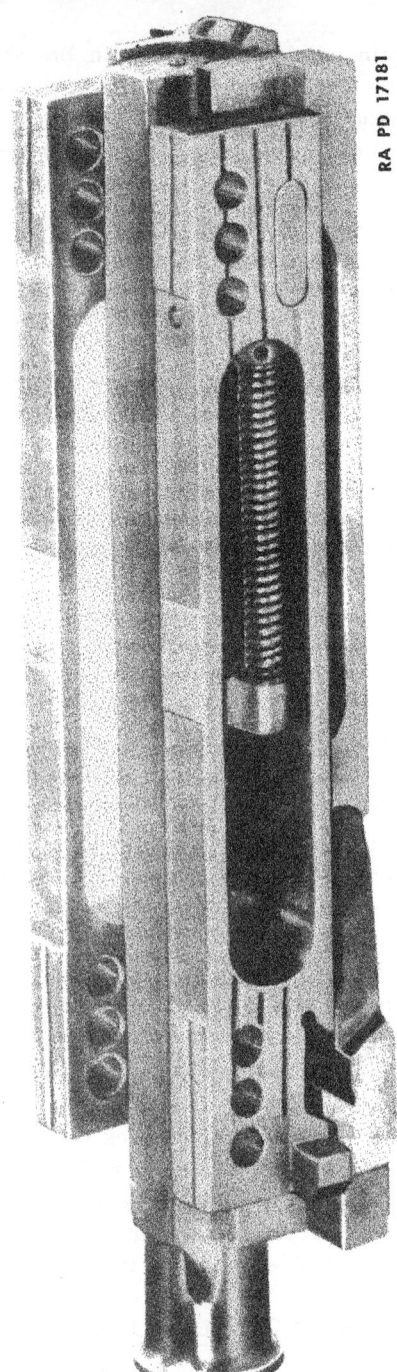

Figure 13 — Breechblock Assembly Without Inertia Blocks — Locked Position

RA PD 17181

16

TM 9-227
13

DESCRIPTION AND FUNCTIONING

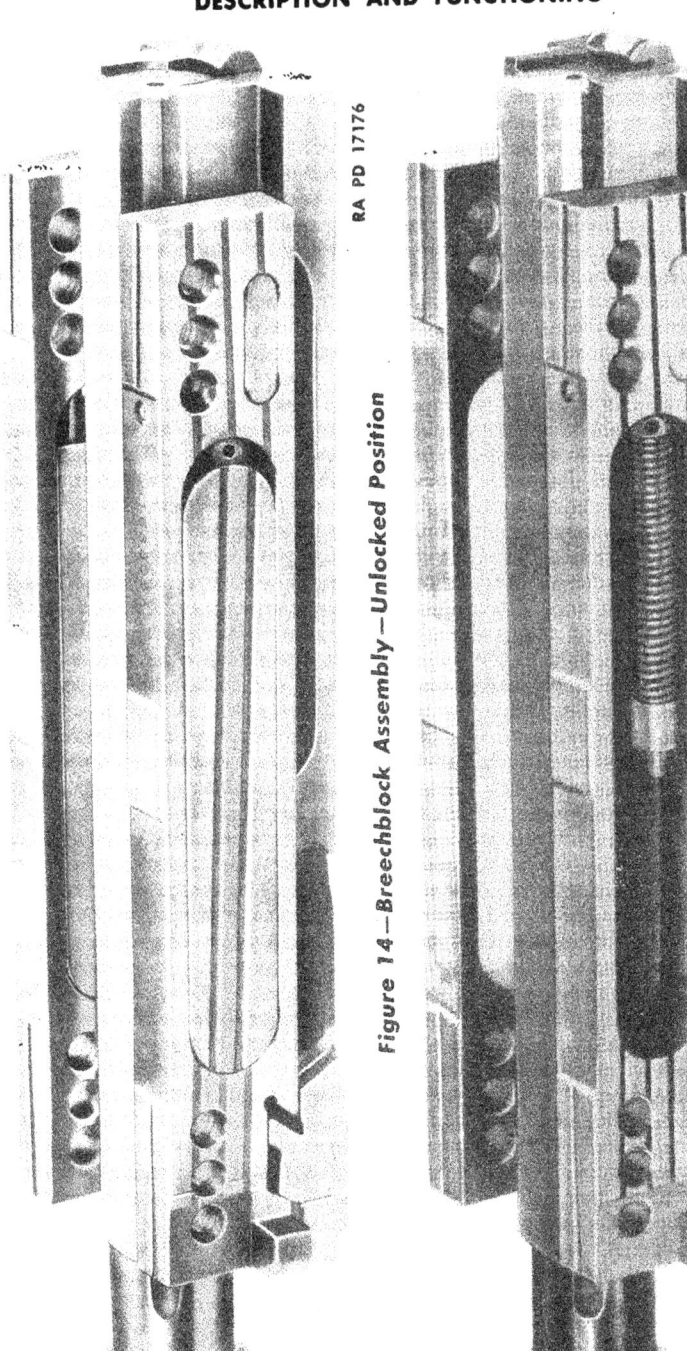

Figure 14 — Breechblock Assembly — Unlocked Position

Figure 15 — Breechblock Assembly Without Inertia Blocks — Unlocked Position

TM 9-227
13

**20-MM AUTOMATIC GUN M1 AND
20-MM AIRCRAFT AUTOMATIC GUN AN-M2**

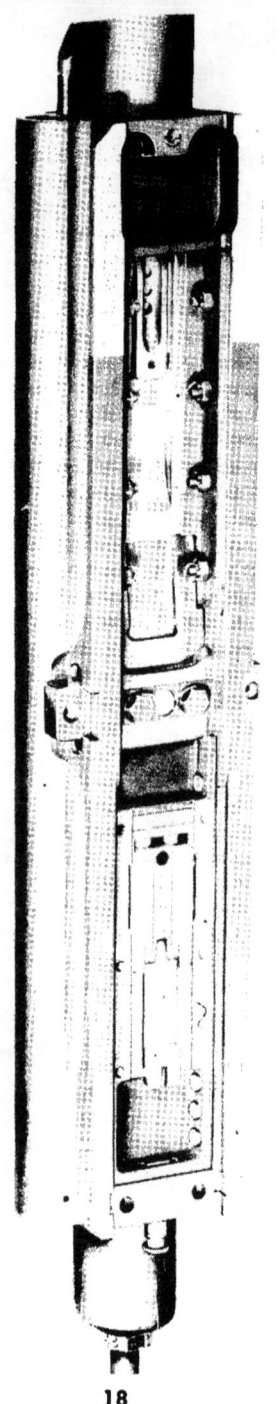

RA PD 10806

Figure 16—Bottom of Receiver Without Sear Cover Plate Group

18

DESCRIPTION AND FUNCTIONING

positioned between the sides of the receiver body and secured to the key by screws. The top of the front face of the key is beveled to allow the breechblock lock to hinge down and engage in front of it. Below the bevel are two projections which support the breechblock lock when it drops to the locked position. The ends of the key project beyond the receiver body and are drilled for attaching the gun to the rear mounting in some installations.

14. SEAR BLOCK GROUP.

a. The sear block group consists of the sear and sear block, together with sear buffer springs, plungers, and blocks (fig. 50). The sear is hinged to the rear of the sear block by a pin. The rear of the sear is forked for engagement with the bowden connection shaft. The sear is operated by the shaft, and its function is to retain the breechblock in the cocked position by engaging the recess in the bottom of the breechblock lock. The sear block is drilled through the front to house the two springs and plungers. The sear buffer blocks, one of steel and one of fiber, provide a front abutment for the sear buffer springs and plungers. The steel block should be adjacent to the plungers; the flat surface of the fiber washer should be adjacent to the steel block. The function of the sear buffer springs is to absorb the shock when the sear and breechblock engage. The vertical hole near the front of the sear block is for inserting the sear block disassembling tool. The tool engages the circumferential grooves on the plungers and holds the springs under compression while the group is removed.

15. SEAR COVER PLATE GROUP.

a. The sear cover plate assembly is secured to the receiver plate by means of six screws; the two screws nearest the rear end require lock washers, the other four screws and the sear housing are locked by locking wire (fig. 48).

b. A hardened insert is fitted into a recess on the inside face of the plate. The purpose of this insert is to assure positive engagement of the sear by its camming action against the mating surface of the sear as the latter moves forward against the buffer springs. To the front of the insert, the plate is drilled and tapped to receive the sear spring housing with the spring and plunger. The sear spring forces the sear spring plunger upward against the sear so that the latter can engage the breechblock lock.

c. To the rear of the insert, the plate is drilled and tapped to receive the bowden shaft housing nut. The bowden connection shaft is a shouldered cylindrical shaft which slides vertically within the spring in the

TM 9-227
15

**20-MM AUTOMATIC GUN M1 AND
20-MM AIRCRAFT AUTOMATIC GUN AN-M2**

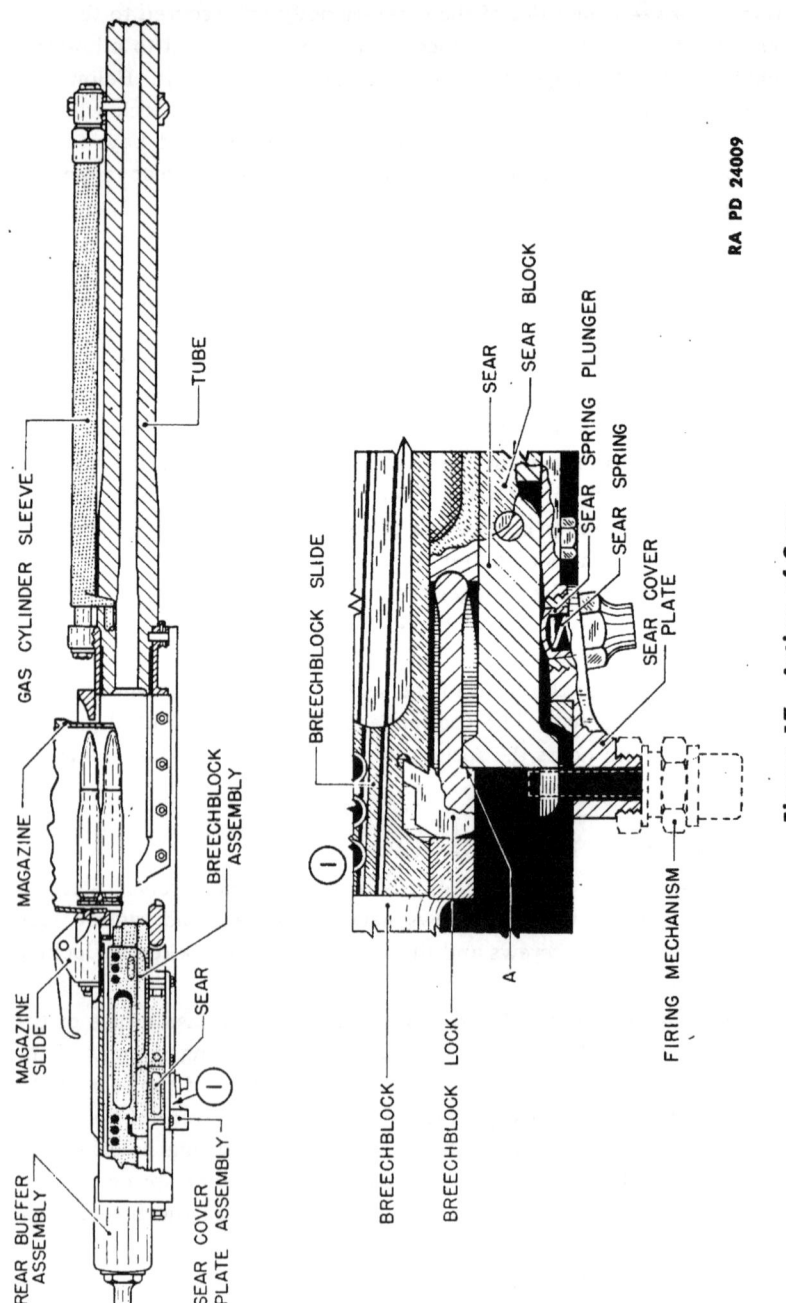

Figure 17 — Action of Sear

DESCRIPTION AND FUNCTIONING

bowden connection nut. The upper end of the shaft has lugs for engaging the forked end of the sear. The lower end of the shaft is drilled and slotted for connecting the bowden control cable. A groove in the shaft is for engagement with the safety trigger pin which is operated by the safety lever. The pin has two notches which, in conjunction with spring and ball, hold the lever in the "SAFE" or "FIRE" position. The lower end of the bowden shaft housing nut is provided with the bowden connection nut which houses the inner and outer bowden connection bushings.

NOTE: In aircraft installations, firing is always done with a solenoid. For this reason, the following components are not used: safety lever mechanism and all attachments protruding from the threaded stud of the sear cover plate proper. The safety lever and the accessory pin, ball, and spring are purposely eliminated to prevent sticking of the shaft.

16. REAR BUFFER ASSEMBLY.

a. The rear buffer assembly is joined to the receiver body by a dovetail connection and a lock plunger assembly which engages a slot in the receiver plate and prevents vertical sliding of the rear buffer (fig. 46). The function of the rear buffer is to stop recoil of the breechblock assembly, cushion the shock, and start it on its forward movement. The rear buffer houses a spring which is placed under initial compression by screwing in the flanged sleeve. Between the spring and the flange of the sleeve is a washer which absorbs the shock when the breechblock is driven to the rear. The rear face of the buffer housing is threaded to receive the driving spring guide head.

b. The rear buffer is provided with a retainer assembly which prevents the driving spring head from unscrewing. The assembly is a washer with a flange and a pin projecting from one face. The pin engages a hole in rear buffer housing, and the flange engages a recess in the housing. In assembling, the rim of the washer is bent over a flat of the head to lock it (fig. 46).

NOTE: At present, the retainer assembly is not provided with all guns.

17. DRIVING SPRING GUIDE GROUP.

a. The driving spring guide group consists of the driving spring, guide, and plunger (fig. 18). The plunger fits into the interior of the bolt and the head rests against the back of the firing pin while the rear end slides in the driving spring guide tube. The driving spring is positioned between the head of the plunger and the head of the driving spring guide. The function of the driving spring is to drive the breechblock group forward, to fire an initial round, and to assist in firing all rounds.

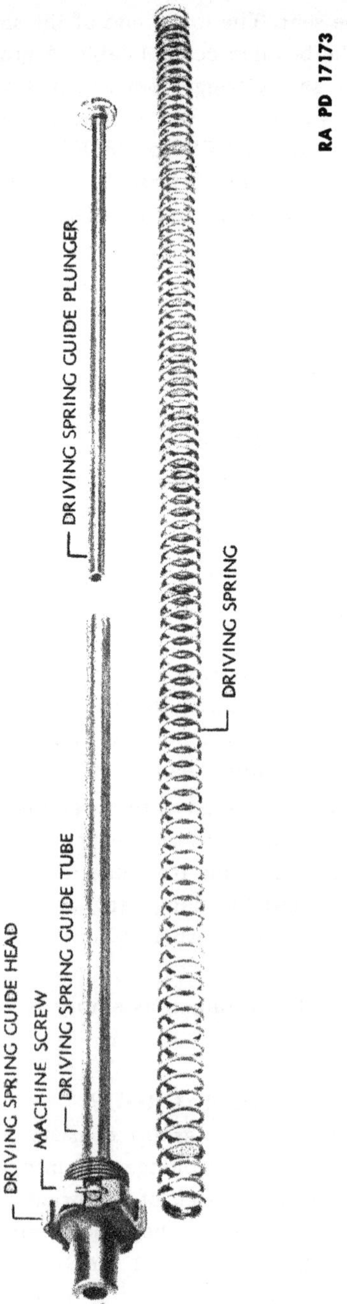

TM 9-227
17

**20-MM AUTOMATIC GUN M1 AND
20-MM AIRCRAFT AUTOMATIC GUN AN-M2**

Figure 18—Driving Spring Guide Group

TM 9-227

DESCRIPTION AND FUNCTIONING

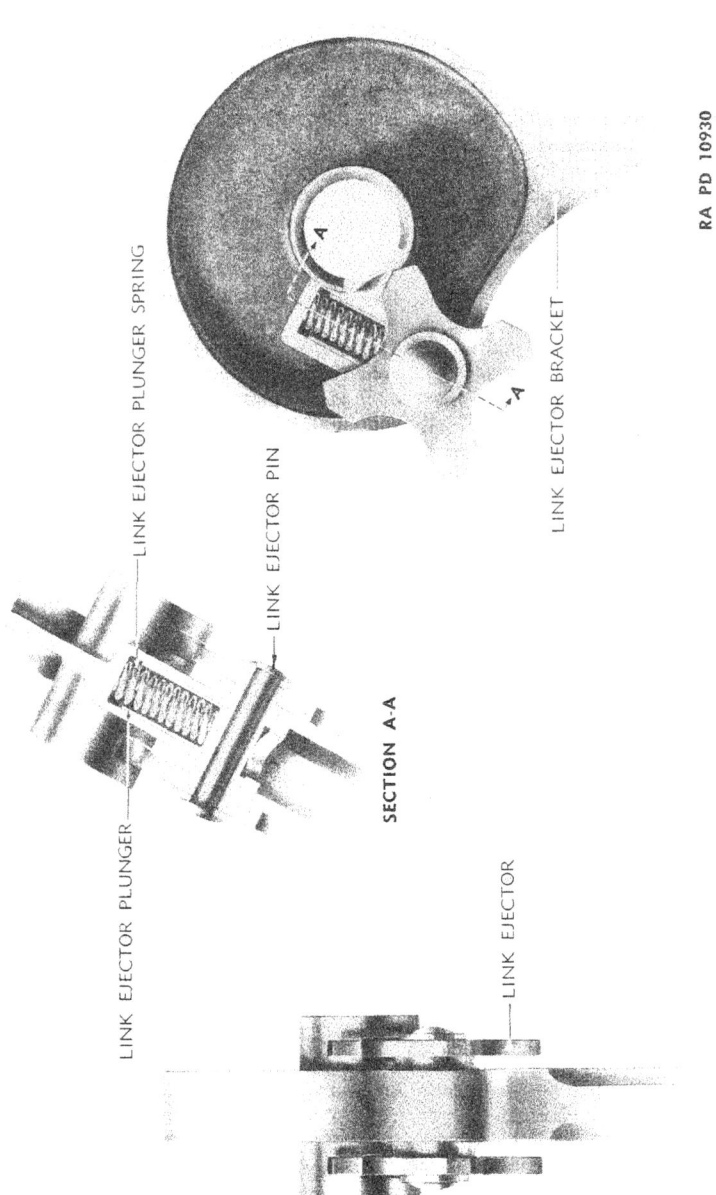

Figure 19 — Link Ejector Assembly

TM 9-227

**20-MM AUTOMATIC GUN M1 AND
20-MM AIRCRAFT AUTOMATIC GUN AN-M2**

18. 20-MM FEED MECHANISM M1.

a. The 20-mm Feed Mechanism M1 comprises essentially a closed cylindrical metal case containing a rotatable central shaft with three sprockets and a spiral driving spring. The front and rear covers are held by three tie rods. The driving spring and the band brake are housed in a case riveted to the front sprocket (figs. 61 and 62). A link ejector (fig. 19) is mounted on the hub of the center sprocket, and a rear feed lever on the rear sprocket. There are two distinct mechanisms: one for right-hand, and one for left-hand feed.

b. The mechanism is operated by the tension of the initially wound driving spring, but the tension is maintained by the recoil of the gun which actuates the rack operating assembly (fig. 60). It takes a recoil of about 20 millimeters (13/16 in.) to operate the feed properly. The rack operating assembly consists of a special gas cylinder guide A25940 which mounts a bracket with an operating lever having an inclined surface at the rear. A roller is pinned to the lever below the inclined surface. When the gun recoils, the lever roller rides up the inclined face on front of the magazine slide while the rack roller rides up the inclined surface on the operating lever. This combined movement actuates the tensioning ratchet to maintain the tension of the driving spring. Unwinding of the spring is prevented by the ratchet tensioning pawl in the front cover. Overwinding is eliminated by the band brake.

c. The belt enters the feed mechanism through the belt guide and the rounds are engaged and carried around by the sprockets. The cartridges contact a cam surface in the front cover and are pushed to the rear, releasing the rounds from the links. The rounds engage the link ejector, causing it to rotate, and causing the points on the ejector to engage the middle freed link and push it outward. The double loops bear against the link chute cover spring, which acts as a fulcrum, and the single loop is moved clear of the next cartridge along with double loops of the next link. The pivotal movement of the link is limited, and reengagement of the single loop with the round is prevented by the outward movement of the double loops of the next link. The links are then forced through the link chute. A spring loaded plunger holds the link ejector in position ready to be engaged by the next round in the belt.

d. **Right- and Left-hand Parts for 20-mm, Feed Mechanism M1.** The description in paragraph 18 b above is applicable for either left- or right-hand feed. Left-hand parts are of the same dimensions and shapes as right-hand parts, but their positions in the feed mechanism are reversed.

DESCRIPTION AND FUNCTIONING

19. 20-MM 60-ROUND MAGAZINE M1.

a. The 20-mm 60-Round Magazine M1 comprises essentially an outer casing closed by front and rear plates and containing a driving spring in a spring casing in the front plate. A hole in the rear plate accommodates the feed arm axis tube. The inner end of the spring is attached to the tensioning tube. On the inside of the plates are spirals which act as guides for the ammunition (figs. 20, 21 and 59).

b. The magazine is operated by the spring tension. Initial tension is applied during assembly (par. 23 a). Further tension is applied progressively during the loading operations. The tensioned spring acts through the tensioning tube, feed arm axis tube, and feed arm to maintain the platform or follower in contact with the last round. Thus a round is always in position in the magazine mouth. As soon as this round is loaded, the next round is brought into position by the spring.

20. FUNCTIONING OF THE GUN AS A WHOLE.

a. The following is an account of a complete firing cycle from the explosion of one propelling charge to the next:

b. At the moment of firing, the projectile starts down the tube, propelled by the expanding gases. The firing pin is in its forward position, having struck the primer of the cartridge. The breechblock is held in its forward position by the action of the breechblock lock. The lock engages the breechblock at point A, figure 22, and bears against surface B, figure 22, of the breechblock key. The breechblock slide engages the lock at point C, figure 22, preventing the lock from being forced upward prematurely.

c. As the projectile moves forward, it passes the gas port (fig. 23). A portion of the expanding gases enters the gas port, passes through the gas cylinder vent plug, enters the gas cylinder, and exerts pressure on the gas cylinder piston. This piston moves rearward, carrying with it the gas cylinder sleeve. The yoke on the rear end of the gas cylinder sleeve engages push rods and carries them rearward. The push rods, in turn, contact the breechblock slides. The slides are connected by the breechblock slide key, which also engages a slot in the bottom of the firing pin. As the breechblock slides are forced rearward by the push rods, the key carries the firing pin rearward. This retraction of the breechblock slides continues until the rear angle of the breechblock slides clears the end of the breechblock lock. At this point when the two angles clear, the blowback action on the breechblock assembly forces the breechblock lock out of engagement with the breechblock locking key, thus completing the

TM 9-227

20-MM AUTOMATIC GUN M1 AND 20-MM AIRCRAFT AUTOMATIC GUN AN-M2

RA PD 10804

Figure 20 — Rear View of 60-Round Magazine

RA PD 10805

Figure 21 — Front View of 60-Round Magazine

TM 9-227
20

DESCRIPTION AND FUNCTIONING

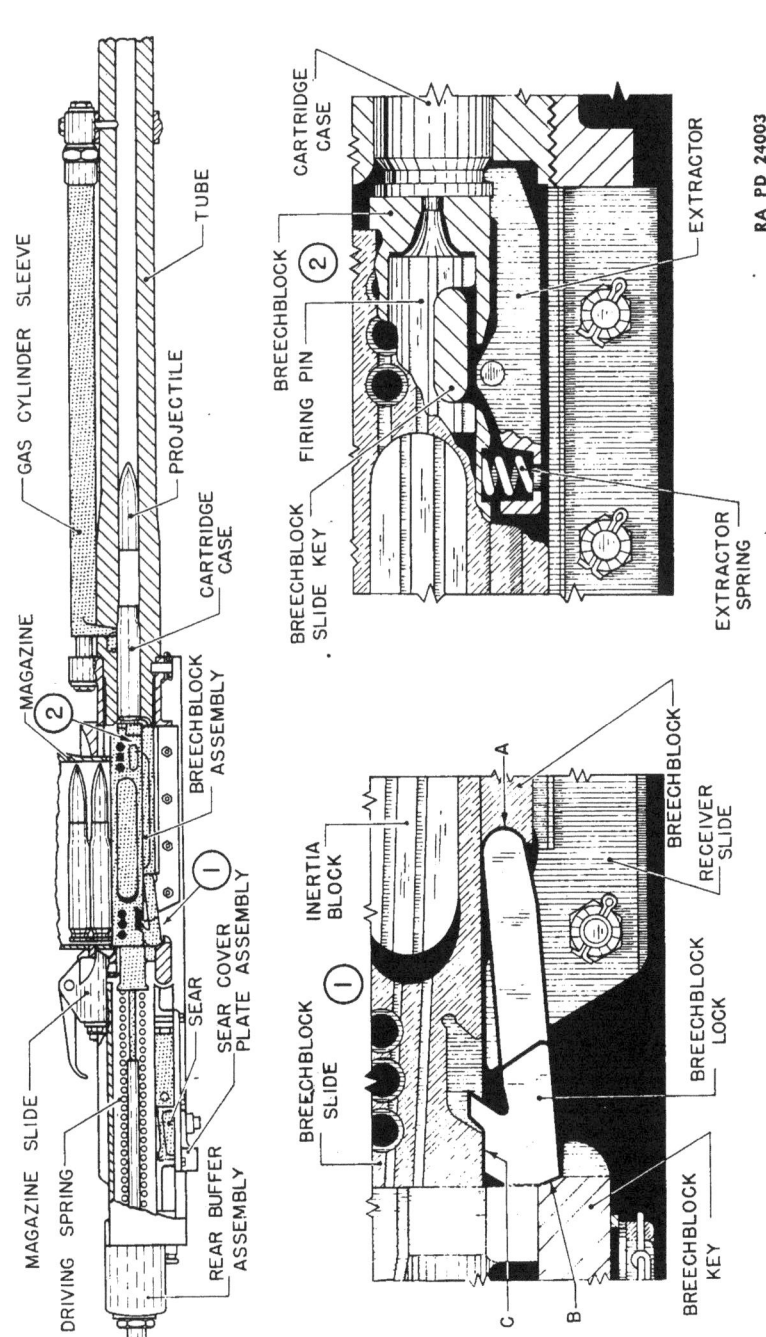

Figure 22 — Gun Mechanism Shortly After Firing

TM 9-227
20-MM AUTOMATIC GUN M1 AND 20-MM AIRCRAFT AUTOMATIC GUN AN-M2

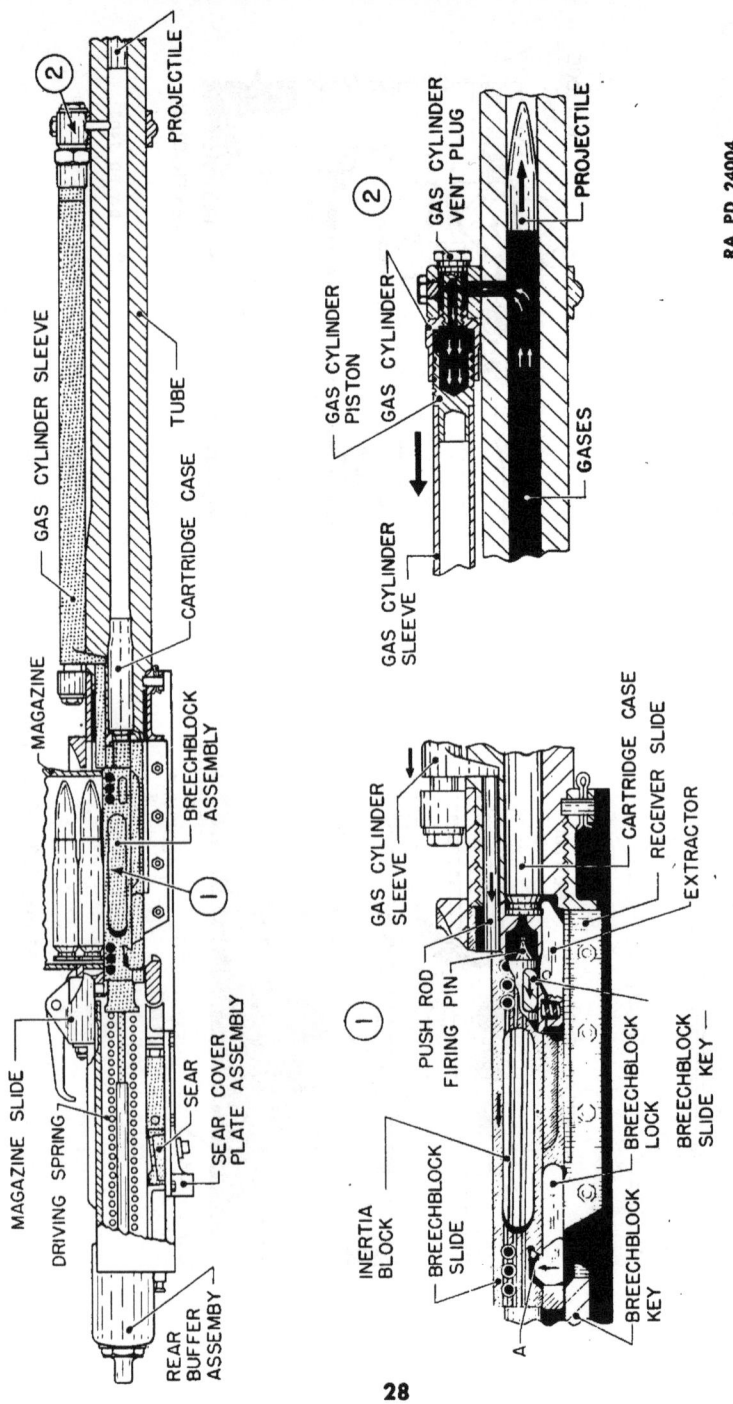

Figure 23—Gun Mechanism During Breechblock Unlocking

DESCRIPTION AND FUNCTIONING

unlocking action of the gun. When the lock has been brought to a horizontal position the breechblock is forced to the rear by direct blowback.

(1) When the breechblock assembly is moved to the rear, the empty cartridge case which has been forcing the bolt back by blowback action is contacted on the upper edge by the two prongs of the ejector (fig. 24), forcing the cartridge case to pivot about and force downward the forward end of the extractor. The cartridge case leaves the hook of the extractor and moves through an opening in the bottom of the gun, completing the ejection of the fired case (fig. 24). When the cartridge case frees itself from the extractor, the extractor is returned to its normal position by the action of the extractor spring.

(2) By this time the gas cylinder sleeve has been returned to its forward position. The gas cylinder sleeve spring, which was compressed during the rearward movement of the sleeve, expands when the gas pressure drops, forcing the sleeve and piston forward.

d. When the breechblock is sufficiently far to the rear to clear the magazine, a new round is forced downward into the lips of the magazine by a spring in the magazine that maintains pressure on the new rounds. In recoiling, the breechblock compresses the driving spring. As the breechblock nears the end of its blowback, it strikes the rear buffer and compresses the buffer spring, which absorbs the remaining force of recoil and brings the breechblock to a stop (fig. 25). As the breechblock comes to a stop, the inertia blocks continue to move rearward in their slots in the breechblock slides until they reach the end of the slots. By this time the breechblock has started forward again and the inertia blocks remain in a rearward position with respect to the breechblock slides during the block's forward motion.

(1) The function of the sear will be described later, but it should be noted at this point that it is held in a downward position during automatic fire, allowing the breechblock to move through its cycle without being caught and held at the rear.

e. The rear buffer spring and the driving spring expand, forcing the breechblock forward (fig. 26). The top of the front surface of the breechblock engages the new cartridge which has been positioned on the lips of the magazine. As the cartridge is forced forward, it drops into the recess in the bolt where it is gripped by the hook on the extractor as it enters the chamber. Note that the inertia blocks are in a rearward position with respect to the breechblock slides.

f. As the breechblock reaches the end of its forward motion (fig. 27), it seats against the end of the tube, closing the chamber. The momentum of the slides causes them to continue to move forward, releasing the

TM 9-227
20-MM AUTOMATIC GUN M1 AND 20-MM AIRCRAFT AUTOMATIC GUN AN-M2

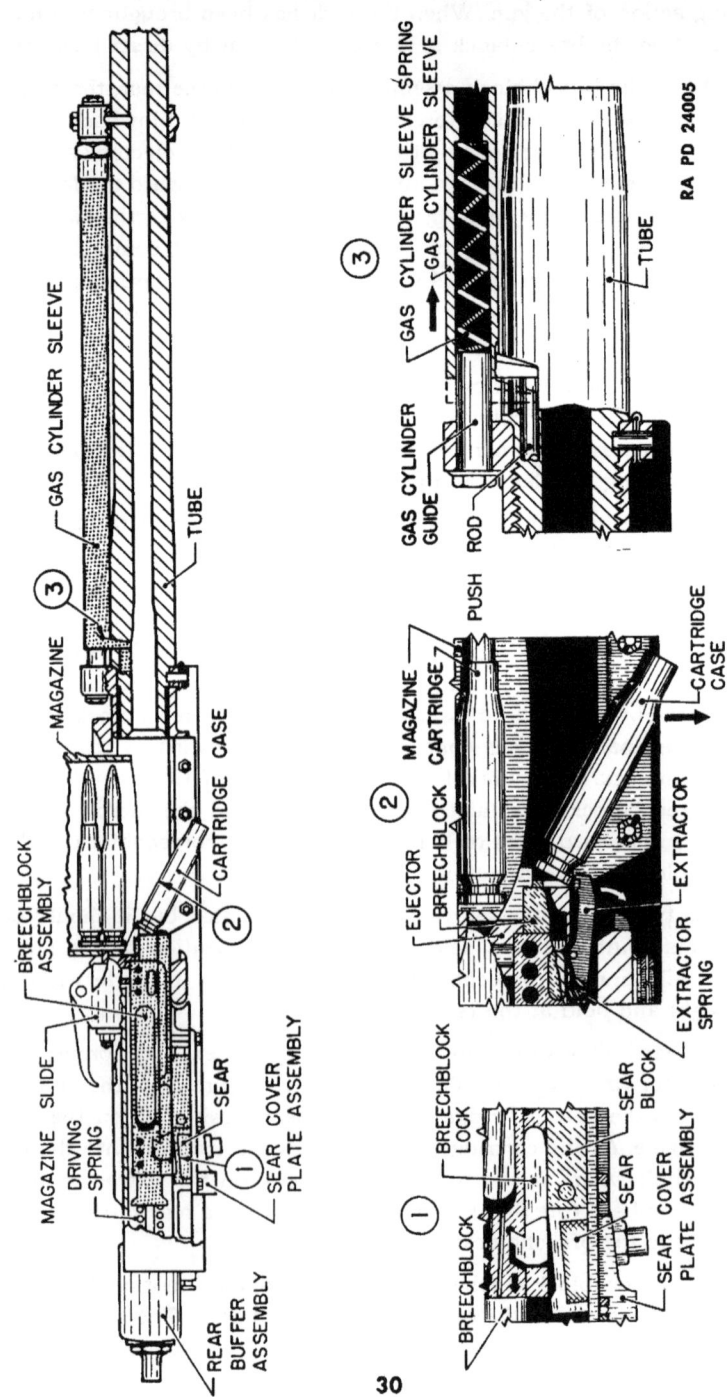

Figure 24 — Gun Mechanism, Cartridge Being Ejected

DESCRIPTION AND FUNCTIONING

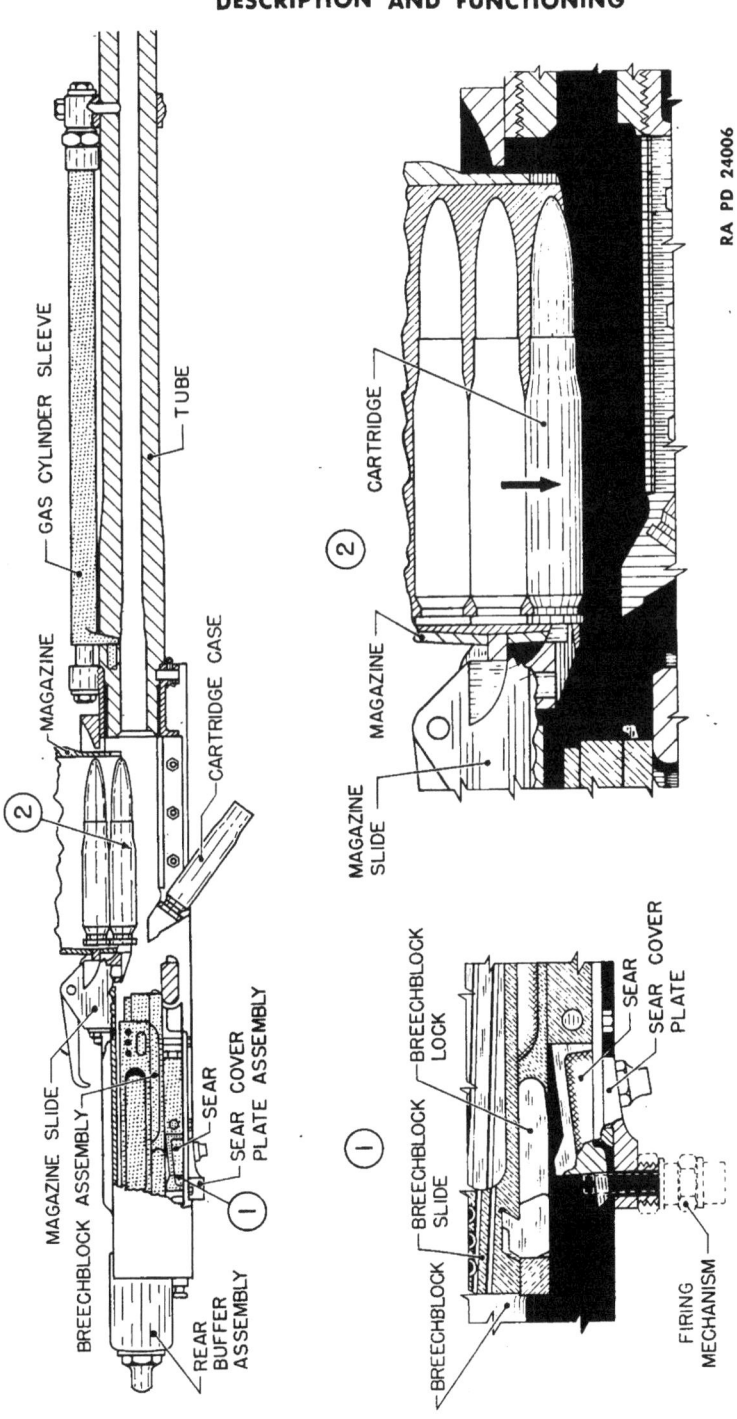

TM 9-227
20

Figure 25 — Gun Mechanism, Breechblock in Rearmost Position

TM 9-227
20

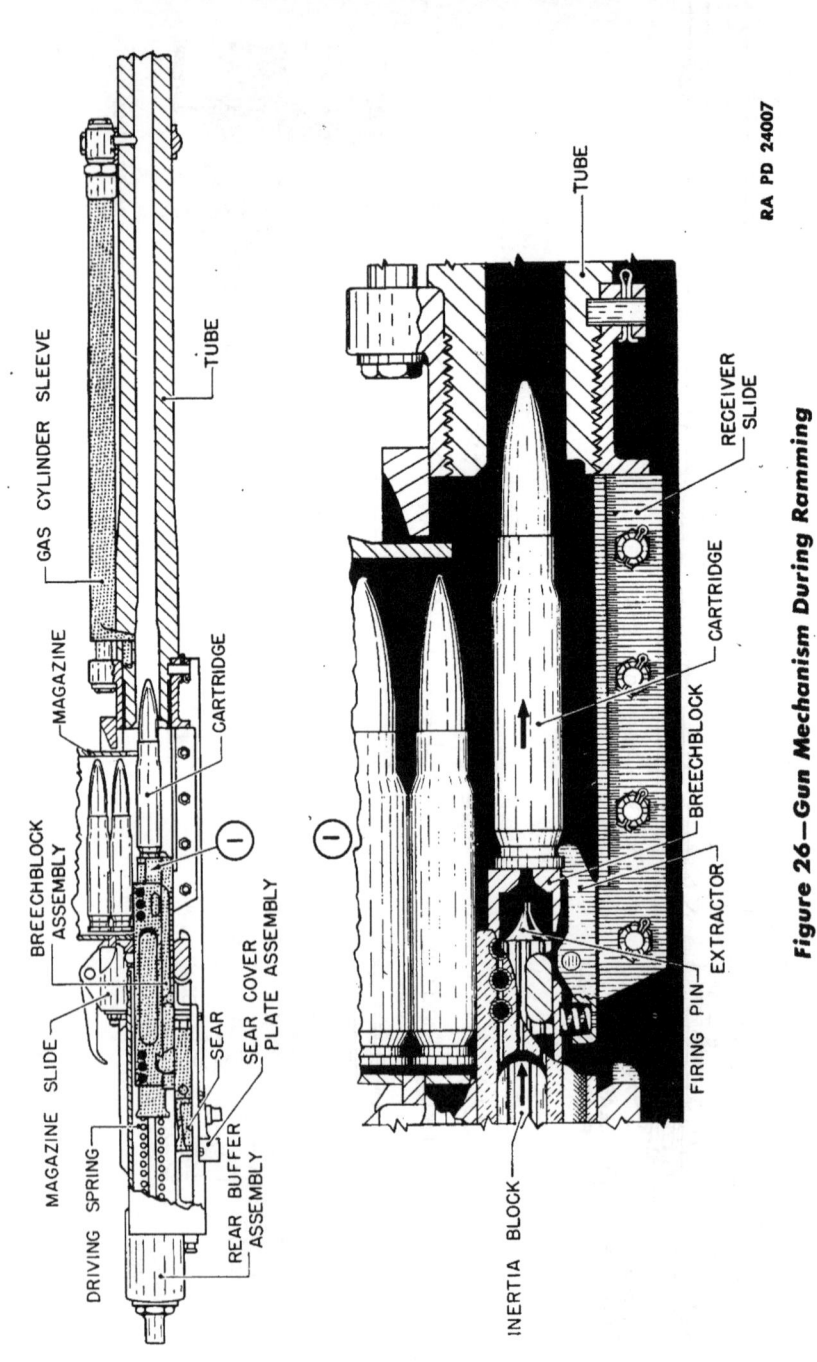

**20-MM AUTOMATIC GUN M1 AND
20-MM AIRCRAFT AUTOMATIC GUN AN-M2**

Figure 26 — Gun Mechanism During Ramming

DESCRIPTION AND FUNCTIONING

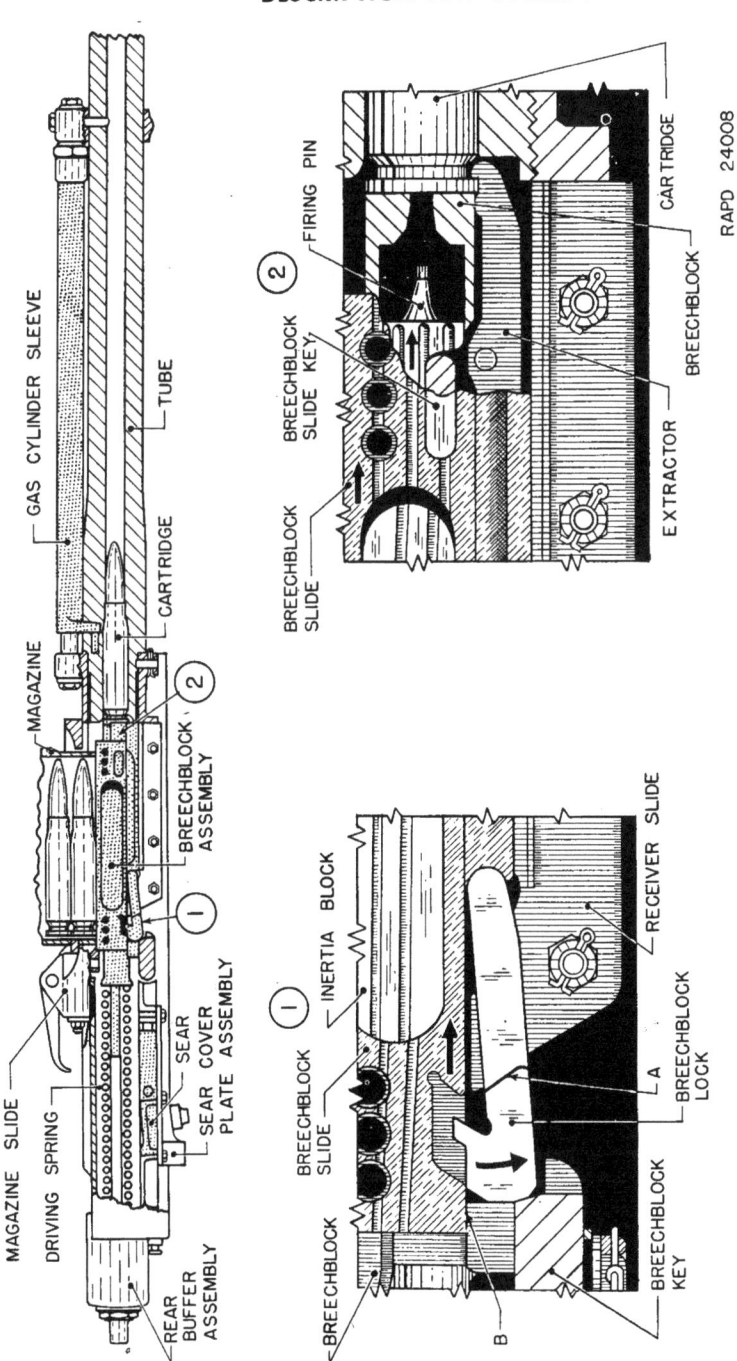

Figure 27 — Gun Mechanism Ready to Fire

20-MM AUTOMATIC GUN M1 AND
20-MM AIRCRAFT AUTOMATIC GUN AN-M2

breechblock lock. At the same time, cam surfaces (A, fig. 27) of the lock are engaged by projections on the receiver slides and the came surfaces at the rear of the breechblock notches, forcing the lock downward. The lock seats against the breechblock key and is locked in its downward position by the lower surface (B, fig. 27) of the slides moving over the end of the lock. When the slides strike the end of the tube, the inertia blocks continue to move forward for a short distance, counteracting any tendency that the slides might have to rebound and unlock the breechblock. The breechblock slides, in continuing forward, carry the firing pin with them. The firing pin strikes the primer of the cartridge, firing it and starting the cycle all over again.

g. To stop firing of the gun, the trigger mechanism is released (fig. 17). The sear is forced upward by the sear spring and plunger. As the breechblock moves rearward, the breechblock forces the sear down. On the block's forward motion, the sear rises again and enters a slot in the breechblock lock, engaging the lock at point A, figure 17, and stopping the forward motion of the breechblock assembly. The shock is taken up by springs and plungers in the sear block.

Section III

OPERATION

	Paragraph
Lubrication of ammunition	21
Loading a belt for 20-mm Feed Mechanism M1	22
Loading the 60-Round Magazine M1	23
Cocking the gun	24
Loading the gun	25
Firing the gun	26
Unloading the gun	27

21. LUBRICATION OF AMMUNITION.

a. Although the use of oil or grease on ammunition is generally prohibited, it is necessary to lubricate the cartridge cases of rounds for these guns. This is done to facilitate extraction and prevent jamming.

b. Dip a cloth in OIL, lubricating, preservative, light. Then wring it out and wipe the curved surface *only* of the cartridge cases with it, applying a thin film of oil.

c. Extreme care must be taken to avoid oiling the primer (base of the case) or the joint where the case is crimped to the projectile.

d. Only one day's supply of ammunition should be lubricated at a time. Rounds oiled for firing and not used the same day should be wiped dry to prevent oil seepage and the accumulation of dirt. Use such rounds first in later firing. Oil them again before use.

e. If OIL, lubricating, preservative, light, is not available, use OIL, lubricating, for aircraft instruments and machine guns.

22. LOADING A BELT FOR 20-MM FEED MECHANISM M1.
a. Right-Hand Feed.

(1) Lay the links along the bench, with their open sides up, double loops to the right, and single loops positioned between the double loops.

(2) The last link at the left end must be of the closed single loop type (fig. 28).

(3) Insert a round into each loop, including the leading double loop, and push it forward.

(4) Check the position of the cartridges relative to the links. The distance from the base of the cartridge case to the back edge of the double loop should be $2^{11}/_{32}$ inches.

(5) Test the belt for flexibility by lifting the left end loop and drawing along the top of the belt to the right. Any faulty link will cause the

20-MM AUTOMATIC GUN M1 AND
20-MM AIRCRAFT AUTOMATIC GUN AN-M2

belt to "kink" instead of folding over smoothly. Any link which does not hinge freely must be replaced by another and the test repeated.

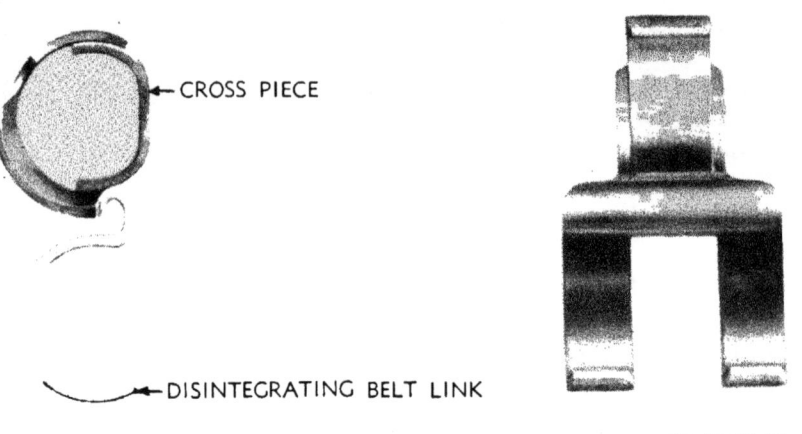

Figure 28—End Link

(6) Repeat the above test, starting with the right end loop and drawing it to the left. If a stiff link is found, it must be replaced by another and both right-end and left-end tests repeated.

(7) Test the belt for oversize links by suspending it from one end and twisting the lowest link until resistance is felt. If the belt breaks, the faulty link must be replaced by another and all tests repeated.

h. Left-Hand Feed. The procedure is the same as in paragraph 22 a above except that the position of the links is reversed. The double loops are to the left, and the special end link is at the right end.

c. Joining a New Belt to a Partly Expended Belt.

(1) Remove the special link from the end of the partly expended belt.

(2) Join the two belts by means of an ordinary link.

(3) Check the position of the round which has been inserted to join the belt.

23. LOADING THE 60-ROUND MAGAZINE M1.

a. If the magazine has been disassembled, apply initial tension as follows:

(1) Place the magazine in the magazine holder or in any other suitable retaining device.

TM 9-227
23

OPERATION

(2) Remove the cotter pin and tensioning tube pin, if necessary.

(3) Rotate the magazine until the follower is in the mouth.

(4) Insert the tensioning tube bar through the end of the tensioning tube and turn it counterclockwise three-quarters of a turn.

(5) Insert the tensioning tube pin and secure it with a cotter pin.

(6) Insert the bar through the hole in the tensioning tube and turn it slightly counterclockwise to lower the platform so as to allow a round to be inserted.

(7) Insert a round, base first, into the mouth of the magazine and push it against the rear plate (fig. 29). Ease the bar, and see that the round is flush against the rear plate (fig. 30).

(8) Turn the bar slightly counterclockwise to lower this first round, and then insert a second round in the same manner.

(9) Repeat this operation until the magazine is full, taking care that the first round inserted contacts the platform or follower. No further tension must be applied to the spring.

Figure 29 — Loading 60-Round Magazine

Figure 30 — Round in Loaded Position

TM 9-227
24-25

**20-MM AUTOMATIC GUN M1 AND
20-MM AIRCRAFT AUTOMATIC GUN AN-M2**

24. COCKING THE GUN.

a. Two types of chargers are under consideration: hydraulic and manual. Information on the hydraulic charger will be found in Air Corps Technical Order 11-1-21. The manual charger will be found in many installations; the details of the charger vary with the make of the aircraft. The manual charger is simple, and probably no technical instructions will have to be issued to maintain it.

25. LOADING THE GUN.

a. **20-mm Feed Mechanism M1.**

(1) Remove the magazine slide group (par. 37 h). Remove the cotter pin, lock washer, and gas cylinder guide. Screw in the special gas cylinder guide.

(2) Attach the bracket, assembled with operating lever and roller, to the gas cylinder guide, and secure with the taper pin. Adjust the magazine slide so that its engraved lines are $\frac{1}{16}$ inch to the rear of the engraved lines on the receiver. Secure the magazine slide so that it will not recoil during firing.

(3) Hold the empty feed mechanism above the magazine slide with the tensioning ratchet pointing toward the muzzle. Carefully lower the mechanism so that the mouth enters the opening of the magazine slide and the latch plate at rear of the mouth rests on the magazine latch.

RA PD 10809

Figure 31 — Placing 60-Round Magazine in Position

OPERATION

(4) Press down on the front end of the mechanism and push it forward until the transversely projecting pins at the front of the mouth engage the hook-shaped projections at the front of the magazine slide.

(5) Lift the magazine slide lever and engage the magazine latch with the latch plate at the rear of the mouth. If the mechanism is properly secured, the rack roller will be just clear of the bottom of the incline on the operating lever.

(6) Rotate the tensioning shaft until one set of teeth on the sprockets are alined with the belt guide. Insert the loaded belt into the belt guide with the double loop of the link leading and the closed side of the link towards the outside of the gun. Push the belt in as far as possible and turn the tensioning ratchet with a wrench, which, in turn, winds the spring and rotates the sprockets, until the driving spring is fully tensioned. This will take two or three complete turns and slippage of the clutch will occur after the proper tensioning has been attained. Do not relieve tension on the wrench after winding until after the second of a series of two clicks has been heard, since this second click indicates that the proper engagement of the ratchet has been made. If winding is stopped after the first click, the relieved load will be brought to bear upon the operating plunger which is not designed for this purpose and which will force it outward against the incline on the operating lever. During the process of winding the spring, two links will be ejected through the link chute. It is always a good policy to make sure that these two links have been ejected which indicates that initial operation of the feed has been normal.

RA PD 10808

Figure 32 — 60-Round Magazine in Position

TM 9-227
25-27

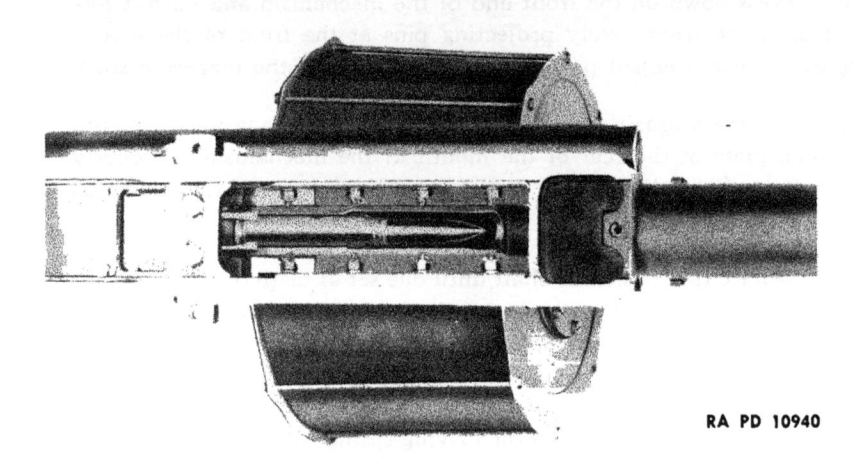

20-MM AUTOMATIC GUN M1 AND
20-MM AIRCRAFT AUTOMATIC GUN AN-M2

RA PD 10940

Figure 33 — Round in Mouth of Magazine

b. **60-Round Magazine M1.**

(1) Place a fully loaded magazine on top of the magazine slide and engage the two pins at the front of the mouth with the hook-shaped projections at the front of the slide (fig. 31).

(2) Lift the magazine slide lever and engage the magazine latch with the rear of the magazine (fig. 32).

26. FIRING THE GUN.

a. **Procedure.**

(1) Cock the gun.

(2) See that the breech is clear.

(3) Operate the firing control mechanism to fire the gun.

b. **To Cease Firing.**

(1) Operate the firing control mechanism to release the sear.

(2) If the sear is not released, the gun will fire automatically as long as live rounds are being fed into the chamber.

(3) In the case of the 20-mm Feed Mechanism M1, all except one round will be expended. The last round of the belt is retained in **the** mechanism.

27. UNLOADING THE GUN.

a. **20-mm 60-Round Magazine M1.**

(1) Point the gun in a safe direction.

40

OPERATION

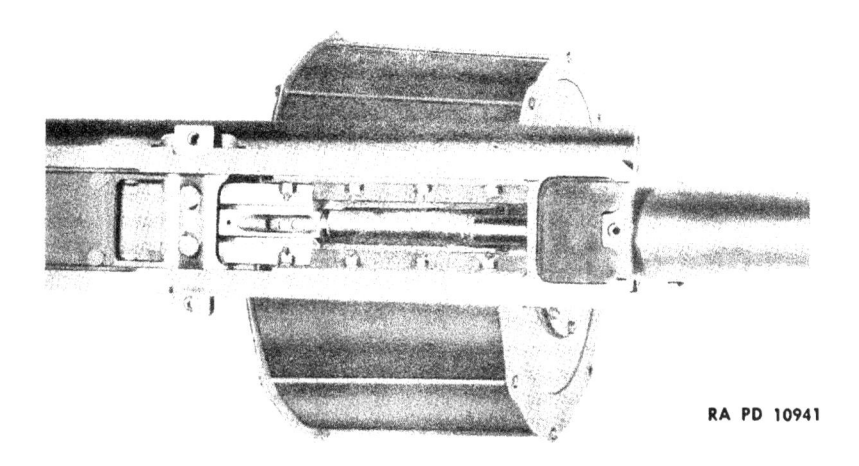

Figure 34—Round Being Pushed by Breechblock Into Chamber

(2) Cock the gun.

(3) Lift the magazine slide lever, disengage the magazine latch from the magazine, then pull the magazine rearward and remove it from the gun.

(4) Fire the gun.

(5) Cock the gun and fire again.

(6) The gun is now unloaded.

b. 20-mm Feed Mechanism M1.

(1) Point the gun in a safe direction.

(2) Cock the gun.

(3) Lift the magazine slide lever and disengage the magazine latch from the feed mechanism.

(4) Pull the feed mechanism rearward and remove it from the gun.

(5) See that the chamber is clear and release the breechblock.

(6) Break the belt near the belt guide by withdrawing a round from the links.

(7) Remove the rounds from the mouth by pushing them forward with a blunt wooden instrument, such as a hammer handle. Do not drop the cartridges as they are removed.

(8) To remove the last round from the mouth, open the link chute cover and, with a screwdriver, push the lower end of the front feed lever so that it rotates and the last round retainer is withdrawn.

(9) Push the last round forward in the mouth and remove it.

TM 9-227
28-29

20-MM AUTOMATIC GUN M1 AND 20-MM AIRCRAFT AUTOMATIC GUN AN-M2

Section IV

MALFUNCTIONS AND IMMEDIATE ACTION

	Paragraph
Immediate action in flight	28
Corrections after flight	29

28. IMMEDIATE ACTION IN FLIGHT.

a. The construction of the gun and its location in the airplane outside the reach of the gunner make it practically impossible to apply immediate action in order to remedy stoppages in flight. The gunner must not attempt to remedy a stoppage in air by recocking the gun and attempting to fire. This may force a new high explosive round against the base of a round in the chamber and cause an explosion.

29. CORRECTIONS AFTER FLIGHT.

a. Immediately after flight, cock the gun. If a hydraulic charging unit is used, leave the pressure on so that there will be no risk of a jammed breechblock moving forward when the pressure is released. Remove the feeder and examine the gun for missing or broken parts.

b. **Failure to Feed.**

(1) See that the magazine carrier is properly adjusted. If the 60-round drum magazine is used, the engraved lines on the magazine slide and receiver must coincide. If the feed mechanism is used, there should be a clearance of from $\frac{1}{32}$ to $\frac{1}{16}$ inch between the roller of the actuating plunger of the feed mechanism and the ramp on the operating link with the feed mechanism loaded and its spring properly wound.

(2) See that all rounds in the 60-round magazine are properly positioned with their bases flush against the rear plate of the magazine. If spring is broken, replace the magazine.

(3) If 20-mm Feed Mechanism M1 is used, check tension of its driving spring. If the recoil of the gun is insufficient to maintain the tension of the driving spring of the feeder, remove the mounting sleeve detent and adjust the compression of the recoil spring by turning the mounting sleeve nut with the special spanner wrench provided (fig. 36). Make this adjustment by actually firing the gun and adjusting the compression until the mechanism winds itself properly.

(4) If the mechanism winds itself properly, see that the cartridges in the links are properly positioned. To clear a jammed link, open the link chute door and remove the link with a screwdriver. Do not insert a

MALFUNCTIONS AND IMMEDIATE ACTION

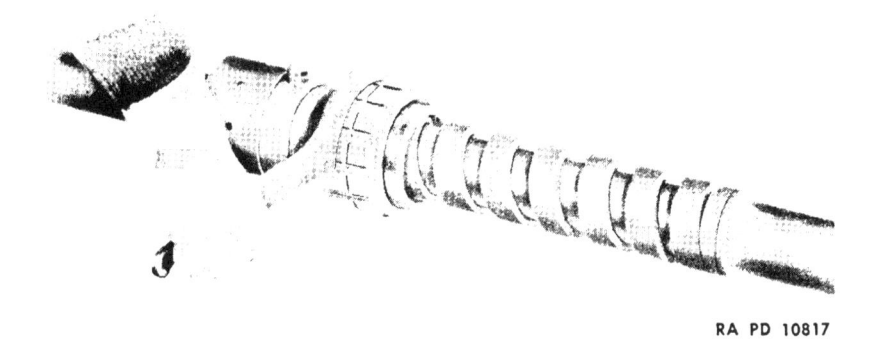

Figure 35 — Adjusting Mounting Sleeve Nut (Front Mounting Not Shown)

finger into the link chute to clear a jammed link because injury may result from the unwinding of the spring.

c. Failure of Round to Enter Chamber in Tube. Examine driving spring; if warped, weak or broken, replace it. Replace the driving spring guide and/or plunger if bent or broken. Tighten the front mounting sleeve nut (fig. 35); see that the magazine slide is secured in correct position with respect to receiver.

d. Failure to Fire. Clean the firing pin hole. If firing pin is short, broken or deformed, replace it.

e. Failure to Unlock. See that the gas cylinder vent plug is clear. Tighten the gas cylinder vent and bracket plugs, and secure with locking wire.

f. Failure to Extract. Replace the extractor and/or spring, if broken. If cartridge case is ruptured in chamber, remove the case.

g. Failure to Eject. Replace the ejector, if broken. See that the magazine slide is secured in correct position with respect to receiver.

h. Run-Away Gun. Replace the sear spring, if weak, broken or missing. NOTE: If the stoppage cannot be remedied by the application of the above listed actions, the gun and feeder should be turned over to ordnance maintenance personnel for inspection and repair.

TM 9-227
30-31

20-MM AUTOMATIC GUN M1 AND
20-MM AIRCRAFT AUTOMATIC GUN AN-M2

Section V

CARE AND PRESERVATION

	Paragraph
General	30
Cleaning instructions	31
Lubrication instructions	32
Replacement of parts	33
Cleaning guns received from storage	34

30. GENERAL.

a. Because of the extremely low temperatures prevailing at all times at high altitudes, special care must be taken to see that aircraft gun lubricants are kept free-flowing. Grease, dried oil, and gum must be very carefully removed before the gun is released for service.

31. CLEANING INSTRUCTIONS.

a. **Gun.**

(1) Before and after each session of firing, the gun should be disassembled and thoroughly cleaned and oiled. In cleaning the bore, use **CLEANER**, rifle bore, applied with a sponge. If this is not available, use a strong solution of issue soap and hot water. Swabbing of the bore should be repeated until a clean flannel patch picks up no foreign matter.

(2) Clean all moving parts with **SOLVENT**, dry-cleaning, and wipe dry with a firm cloth. In cleaning oil cups, open oilholes, and sliding surfaces, do the necessary wiping with a firm cloth. No lint should be allowed to remain in any orifice or sliding parts.

(3) Clean the gas cylinder bracket and the charging cylinder with a brush and then with a flannel patch. Special care should be taken in cleaning the gas port in the tube and the hole in gas cylinder vent plug with wire No. 16, American Wire Gauge (AWG).

(4) All cleaned parts should be examined for wear, scores, burs, and cracks; then oiled and assembled.

b. **Feed Mechanism.**

(1) Before and after each session of firing, wash the feed mechanism or magazine with **SOLVENT**, dry-cleaning, without disassembling. Allow the solvent to drain, and then dry the mechanism. Remove all fouling from the mouth with an oiled rag and wipe dry.

(2) If the magazine is in regular use, disassemble it at least once a month (par. 39). Clean the inside of the magazine with an oiled rag and wipe dry with a clean rag. Take care to remove all dirt from the

CARE AND PRESERVATION

corners of the spirals. Clean the feed arm and follower with an oiled rag and wipe dry.

32. LUBRICATION INSTRUCTIONS.

a. The life of the gun depends to a great extent on proper lubrication. Particular attention should be given to the lubrication of sliding surfaces of the operating mechanism of the gun and other bearing surfaces that do not contain oilholes, plugs, or lubricating fittings.

b. Keep grit out of the lubricant and lubricating openings. In cleaning oil cups, open oilholes, and sliding surfaces, do the necessary wiping with a piece of firm cloth. No lint should be allowed to remain in any orifice or on sliding parts.

c. Lubricate bore, chamber, receiver, and other working parts with OIL, lubricating, preservative, light. If this is not available, use OIL, lubricating, for aircraft instruments and machine guns. When, however, the latter lubricant is used, inspection and lubrication must be made at intervals of not more than 24 hours as this oil has almost no preservative qualities. Apply lubricant with an oiled cloth or oiler after firing or daily while on alert.

d. Oil the feed mechanism or the magazine through the mouth, applying OIL, lubricating, preservative, light, to the working parts. Use no heavy oil or grease.

e. After firing, clean the bore as specified in paragraph 31 a (1) and lubricate with OIL, lubricating, preservative, light.

f. Always lubricate very lightly. Excess oil will impair operation at low temperatures.

33. REPLACEMENT OF PARTS.

a. **Regardless of Condition,** replace parts as follows:
Extractor spring, firing pin, inertia block springs
　(both) . . . . every 1,000 rounds.
Breechblock slides, breechblock slide springs . . . . every 2,000 rounds.
Breechblock lock, extractor, dashpot washer . . every 2,500 rounds.

(1) The driving spring is to be replaced as soon as its free length is less than 23.5 inches.

34. CLEANING GUNS RECEIVED FROM STORAGE.

a. Before placing in service guns received from storage, clean them thoroughly with **SOLVENT**, dry-cleaning, and lubricate as prescribed in paragraph 32.

TM 9-227
35-37

20-MM AUTOMATIC GUN M1 AND
20-MM AIRCRAFT AUTOMATIC GUN AN-M2

SECTION VI

DISASSEMBLY AND ASSEMBLY

	Paragraph
General	35
Special tools	36
Removal of groups and assemblies	37
Disassembly of groups and assemblies	38
Assembly and replacement	39
Disassembly of 60-Round Magazine M1	40
Assembly of 60-Round Magazine M1	41
Disassembly of 20-mm Feed Mechanism M1 (for right-hand feed)	42
Assembly of 20-mm Feed Mechanism M1 (for right-hand feed)	43
Disassembly and assembly of 20-mm Feed Mechanism M1 (for left-hand feed)	44

35. GENERAL.

a. Disassembly and assembly of the gun and accessories should be undertaken only under the supervision of an officer or mechanic. In all cases where the work is beyond the facilities and/or scope of the squadron personnel, ordnance maintenance should be notified in order that qualified personnel with suitable tools and equipment may be provided.

36. SPECIAL TOOLS.

a. Special tools to be used in disassembly and assembly are listed and described in section IX and illustrated in figure 36.

37. REMOVAL OF GROUPS AND ASSEMBLIES.

a. **Driving Spring Guide Group.**

(1) Close the breech. Make certain the breechblock is in its most forward position before proceeding further.

(2) With a blunt chisel or screwdriver, straighten the rim of the retainer so that it does not engage the flat on the driving spring guide head (fig. 37).

NOTE: At present some guns are not provided with retainers.

DISASSEMBLY AND ASSEMBLY

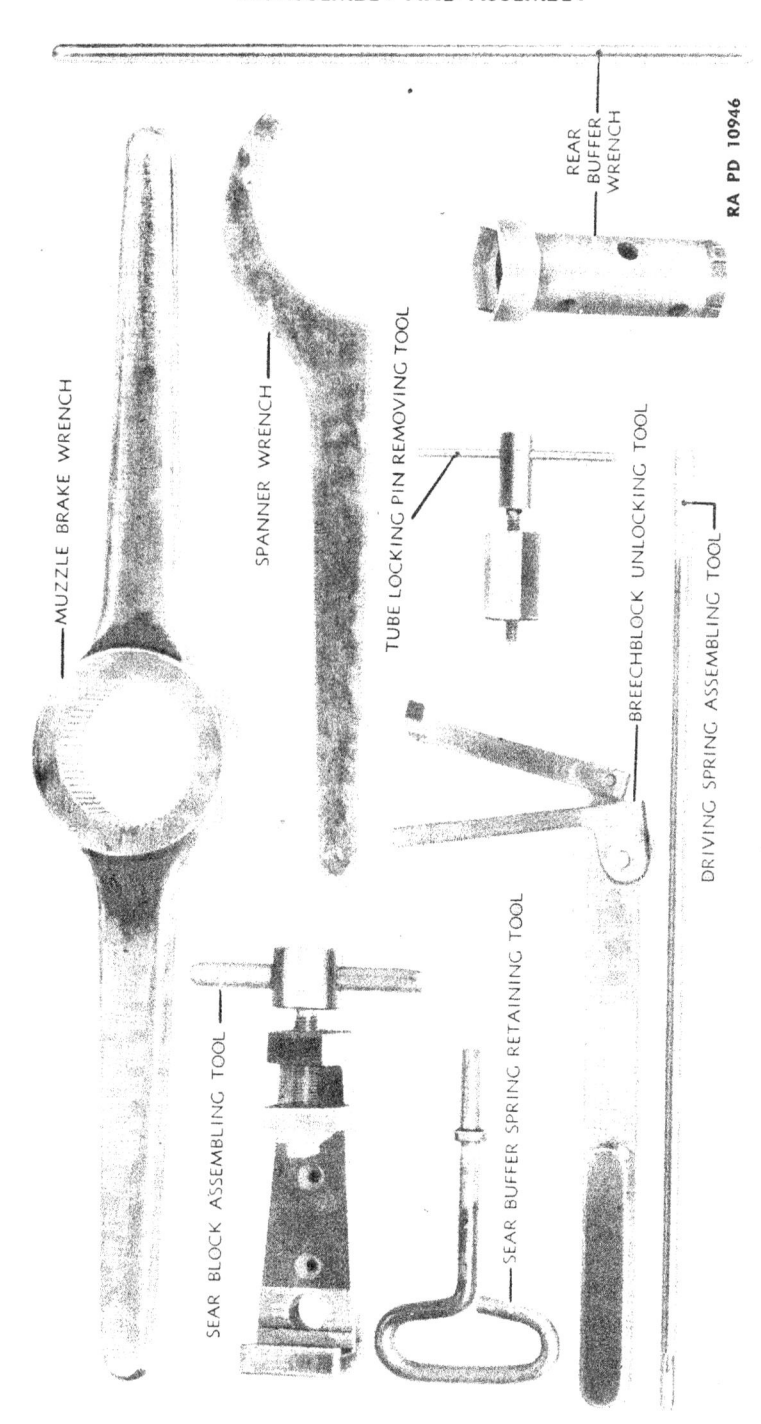

TM 9-227
37

Figure 36 — Special Tools for Disassembly and Assembly

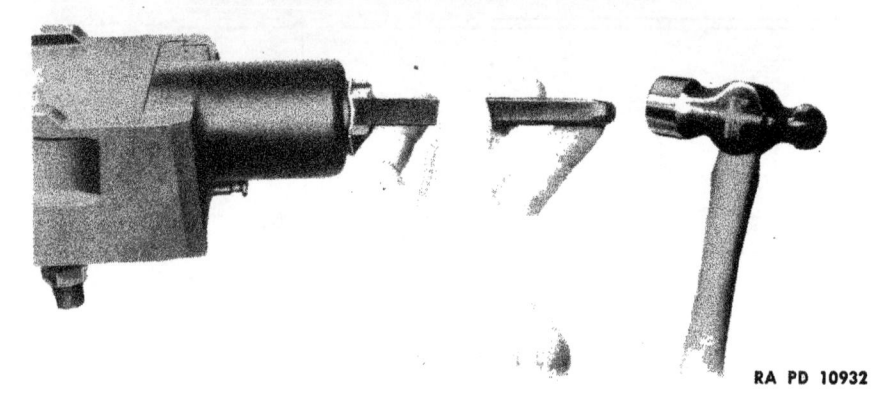

TM 9-227
20-MM AUTOMATIC GUN M1 AND 20-MM AIRCRAFT AUTOMATIC GUN AN-M2

Figure 37 — Straightening the Retainer

(3) Insert the driving spring assembling tool through the driving spring guide head. Push the tool forward until it engages the driving spring guide plunger (fig. 38).

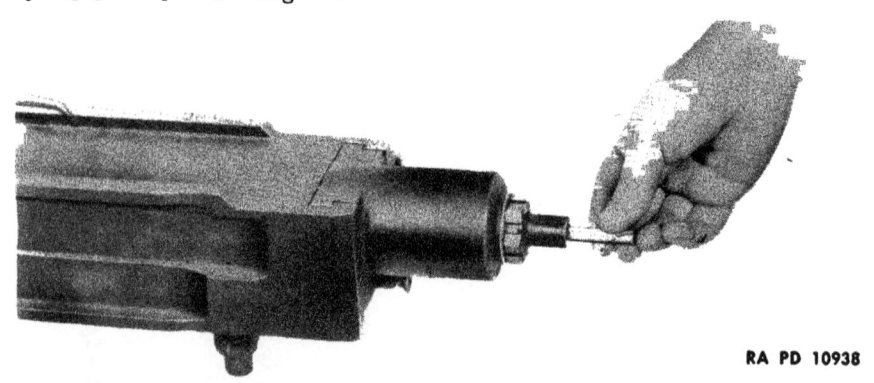

Figure 38 — Inserting the Driving Spring Assembling Tool

(4) Unscrew the driving spring guide assembly, using the rear buffer wrench (fig. 39). Remove the assembly together with the driving spring assembling tool and plunger.

b. Rear Buffer Group. Retract the rear buffer lock plunger (fig. 46), slide the buffer downward and remove it.

c. Breechblock Group.

(1) Engage the projection on the arm of the breechblock unlocking tool with the front face of the right breechblock slide (fig. 40).

(2) Place the other arm of the tool along the top of the breechblock with its end against the receiver (fig. 41).

DISASSEMBLY AND ASSEMBLY

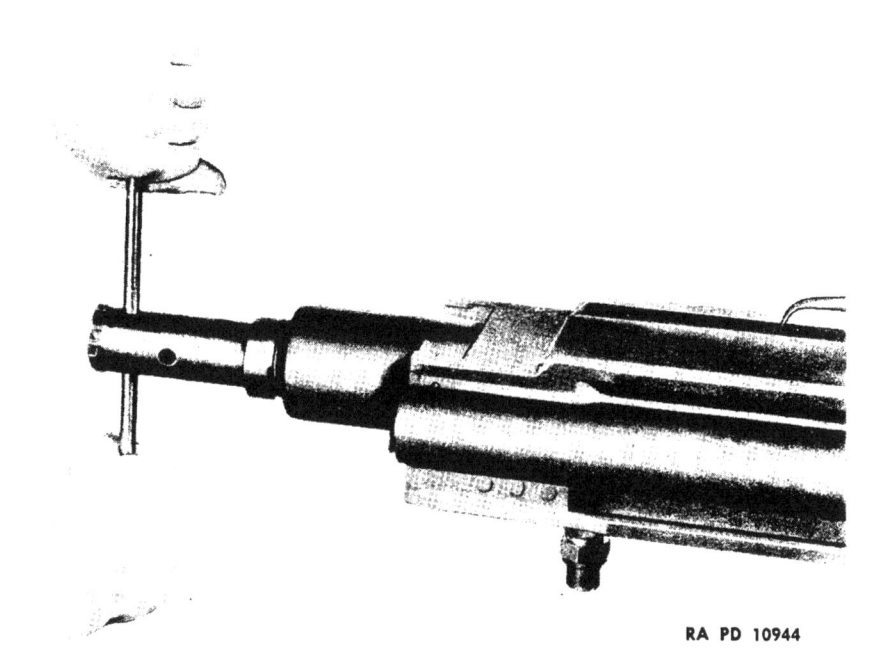

Figure 39 — Unscrewing the Driving Spring Guide Assembly

(3) Press forward the lever of the tool to unlock the breechblock (fig. 42).

(4) Remove the breechblock assembly through the rear of the receiver, taking care to hold the breechblock lock in the unlocked position (fig. 43). Failure to do this may cause the breechblock to get jammed in the rear portion of the receiver as it is being pulled out. Do not drop the breechblock lock.

d. Sear Cover Plate Group.

(1) Place the gun upside down.

(2) Remove the locking wire from the sear cover plate screws and from sear spring housing. Press the plate firmly against the receiver plate and remove the screws. Lift the sear cover plate assembly, taking care not to lose the sear spring and plunger. Keep the gun upside down.

e. Sear Block Group.

(1) Insert the sear buffer spring retaining tool into the hole in the sear block. Push the tool through the sear block so that it fully engages the circumferential grooves on the sear buffer spring plungers.

TM 9-227
37

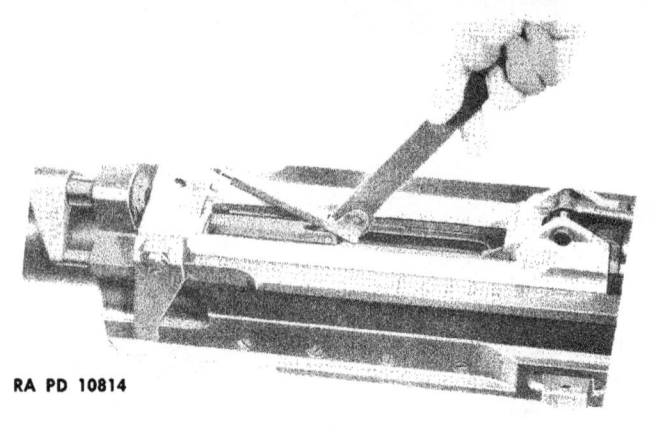

20-MM AUTOMATIC GUN M1 AND
20-MM AIRCRAFT AUTOMATIC GUN AN-M2

RA PD 10814

Figure 40 — Placing the Breechblock Unlocking Tool in Position

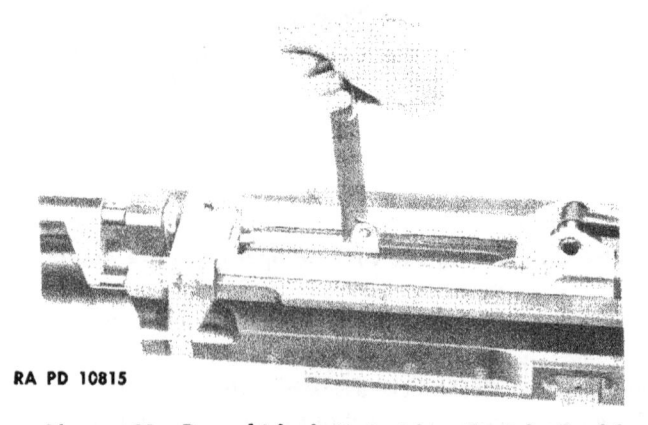

RA PD 10815

Figure 41 — Breechblock Unlocking Tool in Position

RA PD 10816

Figure 42 — Unlocking the Breechblock

50

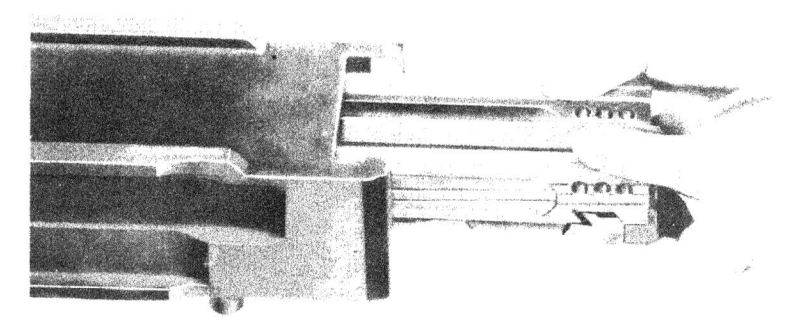

TM 9-227
37-38

DISASSEMBLY AND ASSEMBLY

RA PD 10818

Figure 43 — Removing (or Replacing) the Breechblock

(2) Carefully lift the sear block and sear out of the receiver (fig. 44). Remove the steel and fiber sear buffer blocks.

f. Muzzle Brake Group. Unscrew the muzzle brake assembly, using the special muzzle brake wrench (fig. 45).

38. DISASSEMBLY OF GROUPS AND ASSEMBLIES.

a. Driving Spring Guide Group.

(1) Remove the driving spring guide plunger and driving spring (fig. 18). Withdraw the driving spring assembling tool.

RA PD 10933

Figure 44 — Removing Sear Block Group from Receiver Plate

TM 9-227
38

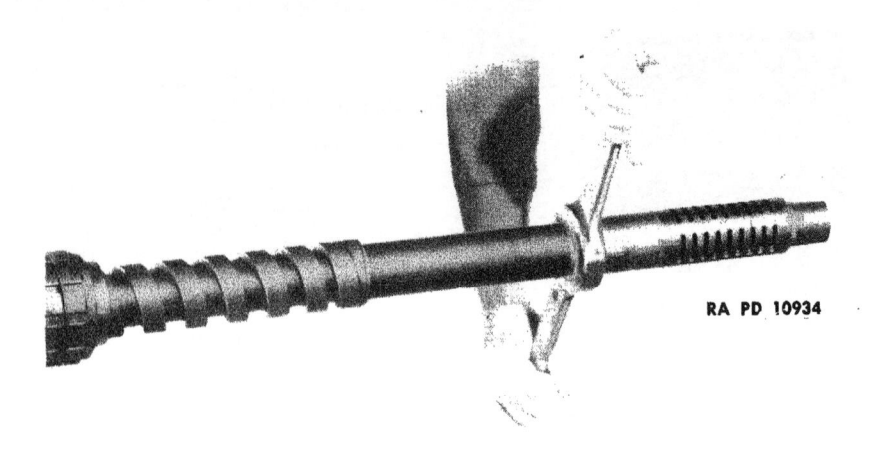

**20-MM AUTOMATIC GUN M1 AND
20-MM AIRCRAFT AUTOMATIC GUN AN-M2**

Figure 45 — Unscrewing the Muzzle Brake Assembly

(2) The driving spring guide and head are attached by a staked screw and a sweated joint and should not be disassembled except by ordnance personnel.

b. Rear Buffer Group.

(1) Drift out the rear buffer lock plunger pin. Remove the plunger, spring, collar, and bushing (fig. 46).

(2) Tighten the rear buffer assembly in a vise with soft jaws. Unscrew the rear buffer sleeve with the rear buffer wrench (fig. 49). Remove the rear buffer washer and spring (fig. 46).

c. Breechblock Group.

(1) Remove the breechblock lock. Remove the left and right inertia blocks (fig. 47). Drift out the inertia block plunger retaining pins and remove the inertia block plungers and springs (fig. 47).

(2) Remove the breechblock slide springs and guides. Withdraw the left breechblock slide and then the right breechblock slide with the slide key assembled (fig. 47). Do not remove the breechblock slide key except for replacement.

(3) Lift the front end of the bolt assembly and allow the firing pin to slide out through the rear. Do not drop the firing pin. Press the extractor against the extractor spring and remove the extractor pin. Withdraw the extractor and extractor spring (fig. 47).

d. Sear Cover Plate Group.

(1) Remove the sear spring and plunger and unscrew the sear spring housing (fig. 48).

52

TM 9-227
38

DISASSEMBLY AND ASSEMBLY

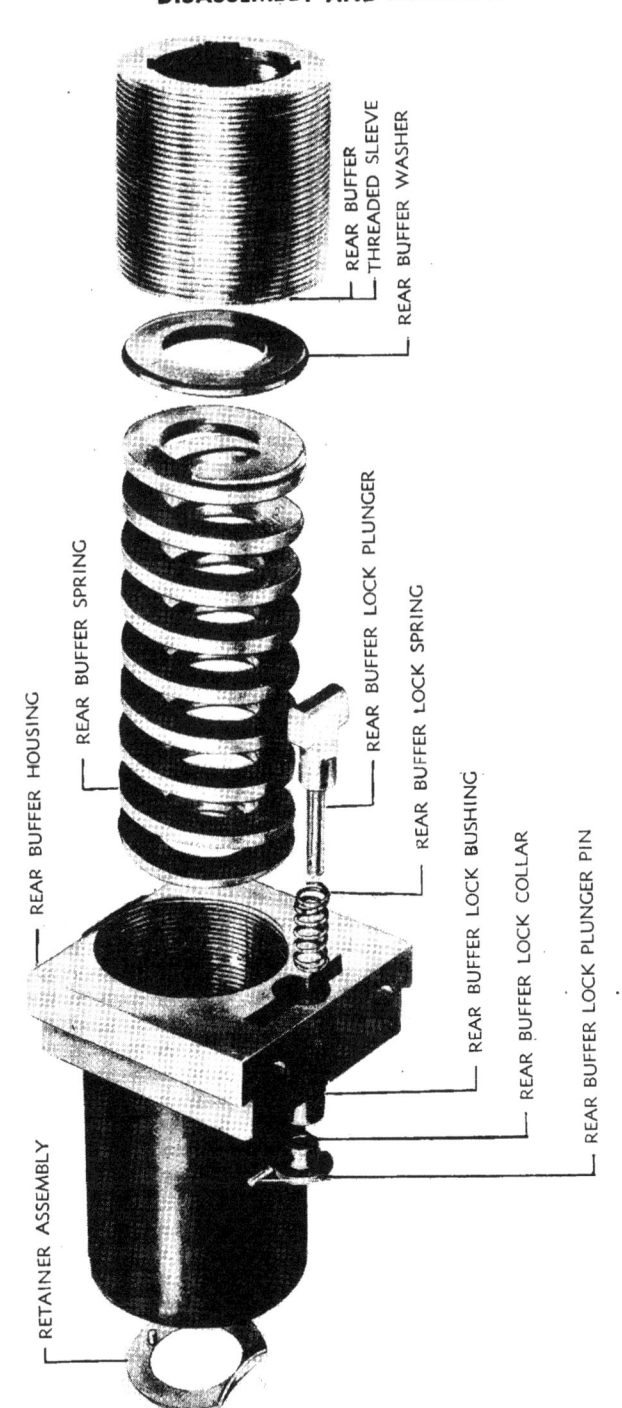

Figure 46—Rear Buffer Assembly

TM 9-227
38

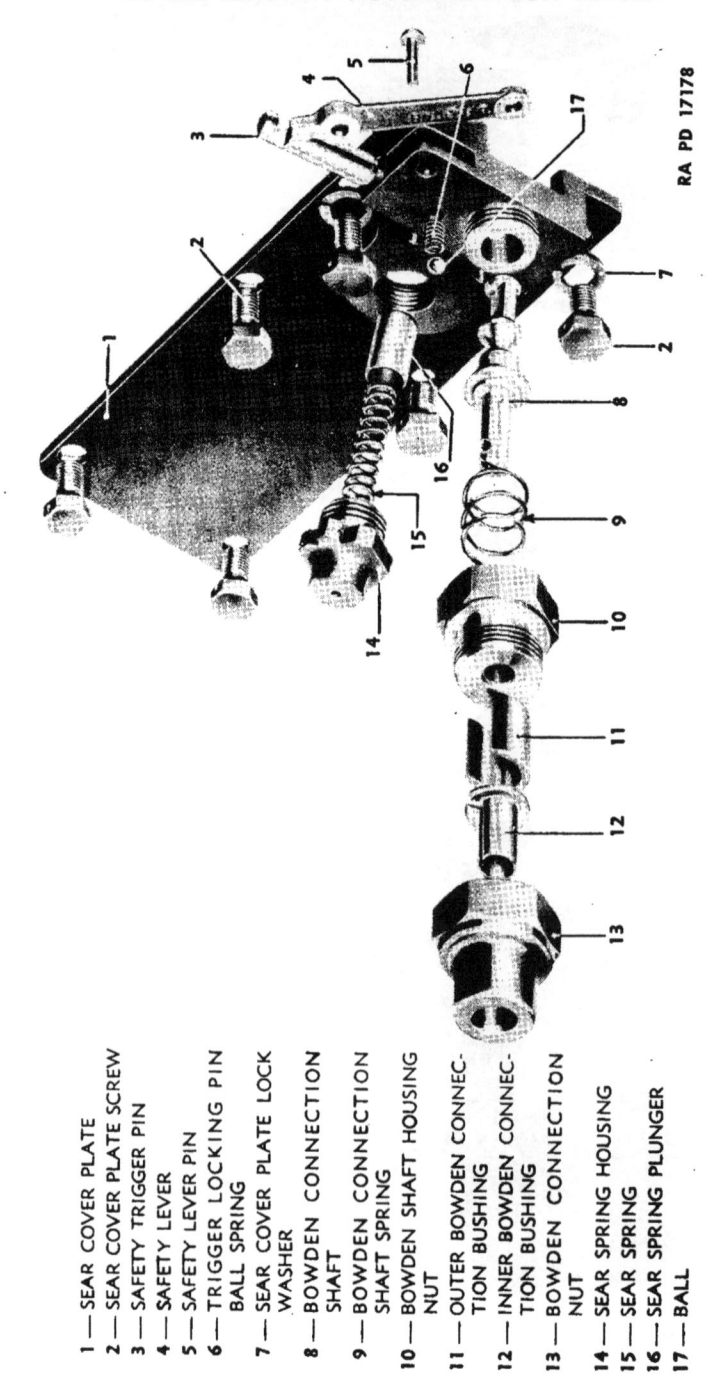

**20-MM AUTOMATIC GUN M1 AND
20-MM AIRCRAFT AUTOMATIC GUN AN-M2**

Figure 47 — Sear Cover Plate Group

1 — SEAR COVER PLATE
2 — SEAR COVER PLATE SCREW
3 — SAFETY TRIGGER PIN
4 — SAFETY LEVER
5 — SAFETY LEVER PIN
6 — TRIGGER LOCKING PIN BALL SPRING
7 — SEAR COVER PLATE LOCK WASHER
8 — BOWDEN CONNECTION SHAFT
9 — BOWDEN CONNECTION SHAFT SPRING
10 — BOWDEN SHAFT HOUSING NUT
11 — OUTER BOWDEN CONNECTION BUSHING
12 — INNER BOWDEN CONNECTION BUSHING
13 — BOWDEN CONNECTION NUT
14 — SEAR SPRING HOUSING
15 — SEAR SPRING
16 — SEAR SPRING PLUNGER
17 — BALL

54

TM 9-227
38

DISASSEMBLY AND ASSEMBLY

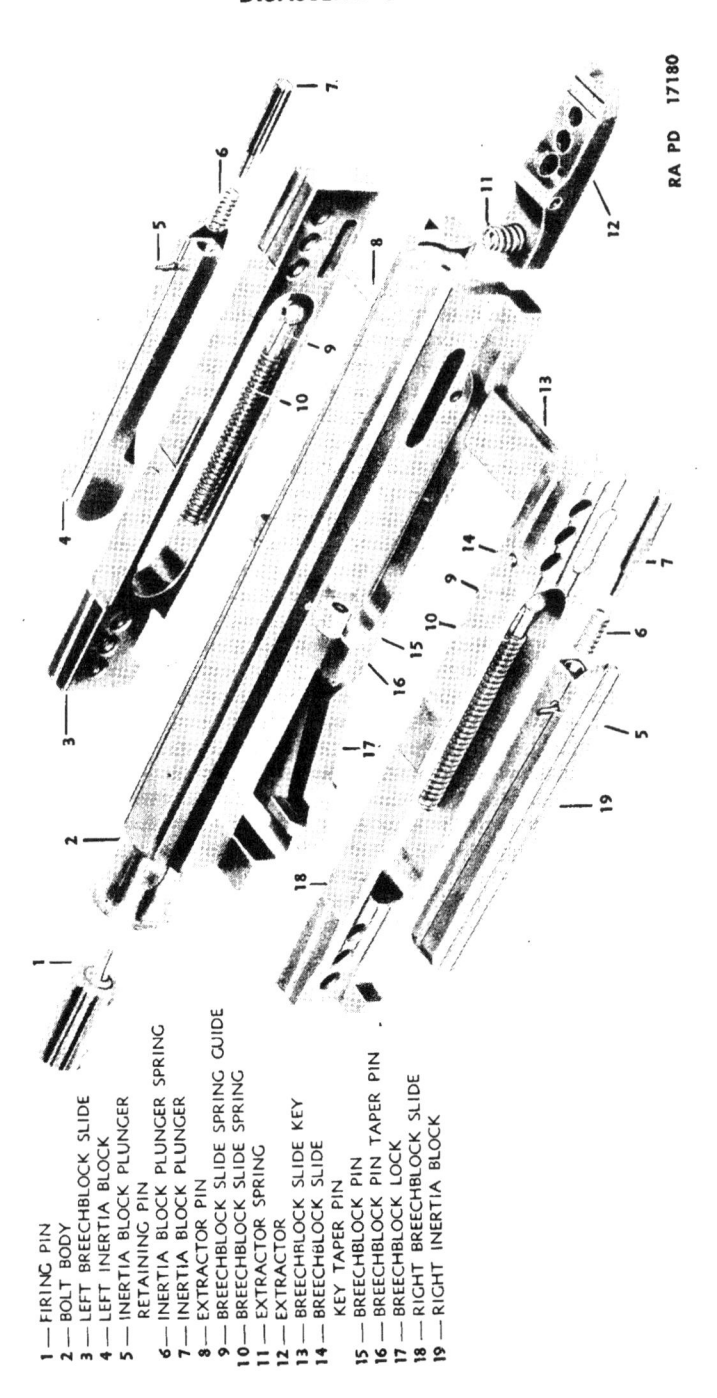

Figure 48 — Breechblock Assembly

RA PD 17180

1 — FIRING PIN
2 — BOLT BODY
3 — LEFT BREECHBLOCK SLIDE
4 — LEFT INERTIA BLOCK
5 — INERTIA BLOCK PLUNGER RETAINING PIN
6 — INERTIA BLOCK PLUNGER SPRING
7 — INERTIA BLOCK PLUNGER
8 — EXTRACTOR PIN
9 — BREECHBLOCK SLIDE SPRING GUIDE
10 — BREECHBLOCK SLIDE SPRING
11 — EXTRACTOR SPRING
12 — EXTRACTOR
13 — BREECHBLOCK SLIDE KEY
14 — BREECHBLOCK SLIDE
15 — KEY TAPER PIN
16 — BREECHBLOCK PIN
17 — BREECHBLOCK PIN TAPER PIN
18 — BREECHBLOCK LOCK
19 — RIGHT BREECHBLOCK SLIDE
20 — RIGHT INERTIA BLOCK

55

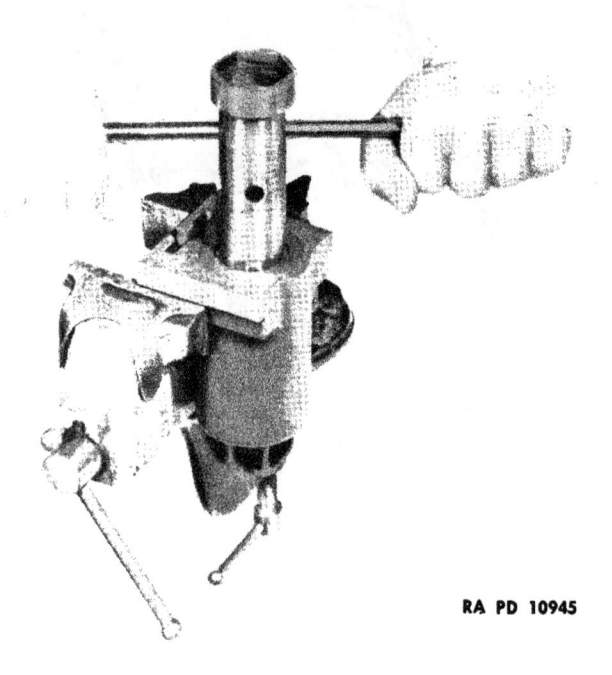

TM 9-227
20-MM AUTOMATIC GUN M1 AND 20-MM AIRCRAFT AUTOMATIC GUN AN-M2

RA PD 10945

Figure 49 — Unscrewing the Rear Buffer Sleeve

(2) Unscrew the bowden connection nut and remove the bowden connection inner and outer bushings. Unscrew the bowden shaft housing nut and remove the bowden connection shaft spring and shaft (fig. 48).

(3) Remove the safety lever pin and safety lever. Withdraw the safety trigger pin and the trigger locking pin ball spring with the ball (fig. 48).

NOTE: In aircraft installations, firing is always done with a solenoid. For this reason the following components are not used: safety lever mechanism and all attachments protruding from the threaded stud of the sear cover plate proper. The safety lever and the accessory pin, ball, and spring are purposely eliminated to prevent sticking of the shaft.

e. **Sear Block Group.**

(1) Withdraw the sear pin to detach the sear from the sear block.

(2) Place the sear block in the sear block assembling tool so that the radial bearing surface of the sear block contacts the jaw while the flat end of the sear block engages the hook-shaped projection at the front of the tool (fig. 42). The sear buffer spring retaining tool should enter the hole in the sear block assembling tool.

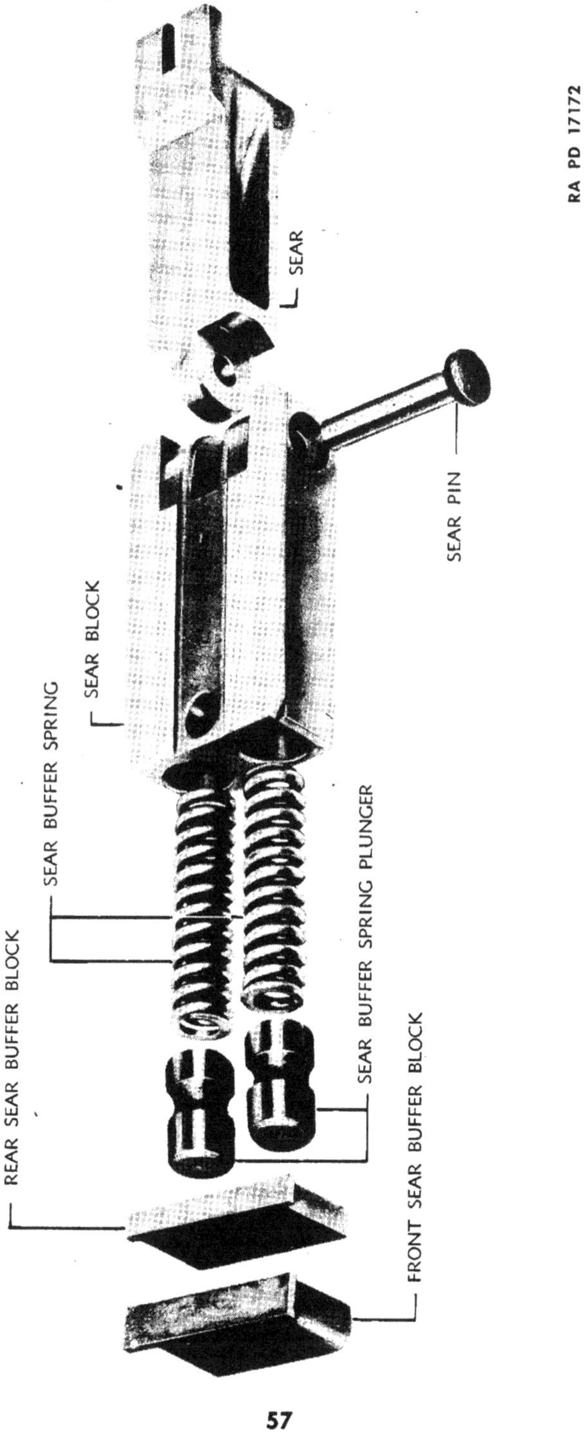

DISASSEMBLY AND ASSEMBLY

TM 9-227
38

RA PD 17172

Figure 50 — Sear Block Group

TM 9-227
38

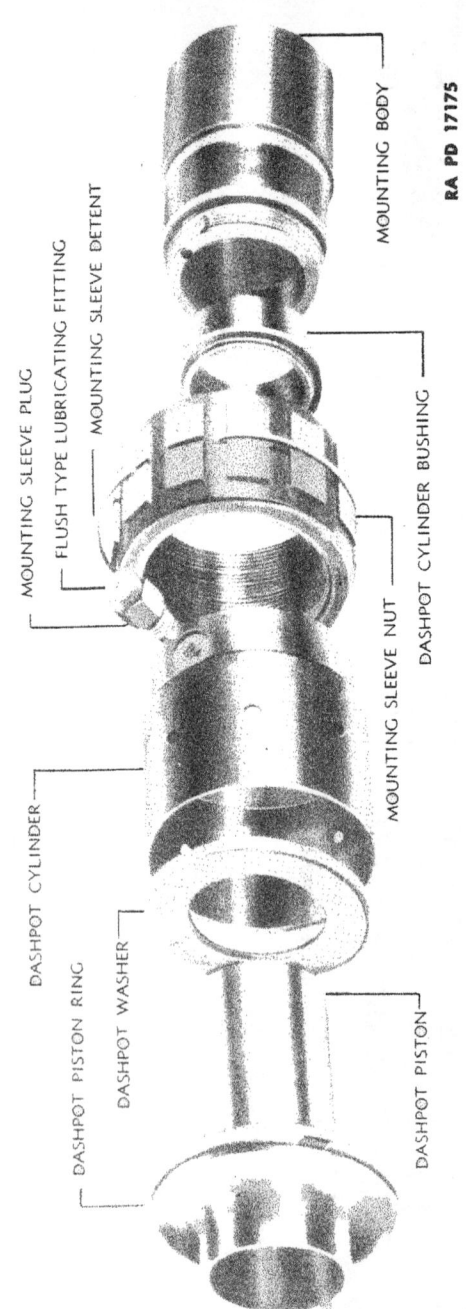

**20-MM AUTOMATIC GUN M1 AND
20-MM AIRCRAFT AUTOMATIC GUN AN-M2**

Figure 51 — Mounting Sleeve Assembly

DISASSEMBLY AND ASSEMBLY

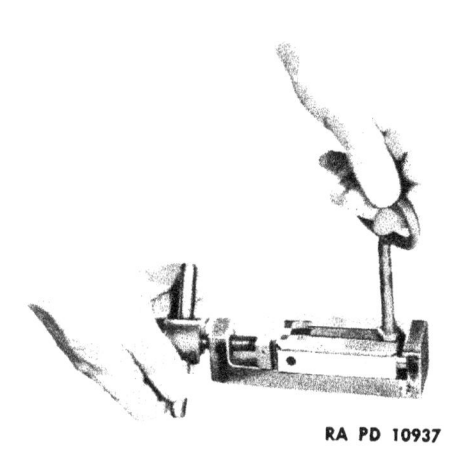

RA PD 10937

Figure 52 — Sear Block Group in Position in the Tool

(3) Turn the handle of the assembling tool sufficiently to take the tension off the retaining tool. Remove the retaining tool (fig. 53). Gradually turn the handle of the tool to release the tension of the springs. Remove the plungers and springs (fig. 50). If the special assembling tool is not available, an ordinary vise will serve. If no retaining tool is available, use a slightly tapered steel rod which nearly fills the hole.

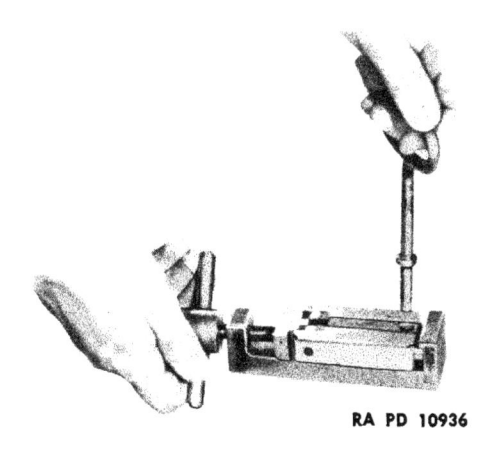

RA PD 10936

Figure 53 — Disassembling the Sear Block Group

TM 9-227
38-39

**20-MM AUTOMATIC GUN M1 AND
20-MM AIRCRAFT AUTOMATIC GUN AN-M2**

f. **Muzzle Brake Assembly.** The disassembly of the muzzle brake assembly is a function of ordnance maintenance personnel only.

g. **Recoil Spring and Mounting Sleeve Group.**
(1) Remove the muzzle brake lock. Slide off the recoil spring sleeve, recoil spring, and the recoil spring filler sleeve (fig. 54).

(2) Slide off the front mounting sleeve assembly and unscrew the plug and fitting. Further disassembly of the mounting sleeve assembly is a function of ordnance maintenance personnel only.

h. **Magazine Slide Group.**
(1) Remove the cotter pin, unscrew the ejector stud nut, and remove the washers and the ejector with the springs (fig. 55).

(2) Remove the locking wire, unscrew the magazine slide backplate screws, and remove the backplate with the magazine latch springs. Remove the magazine slide lever pin and bushing, slide lever, and latch (fig. 55).

(3) Remove the cotter pin and the magazine slide securing arm screw washer, and then the screw. Remove the magazine slide securing arm and slide off the magazine slide (fig. 55).

i. **Gas Cylinder and Sleeve Group.**
(1) Remove the locking wire and unscrew the gas cylinder bracket plug and gas cylinder vent plug. Remove the cotter pin, the lock washer, the gas cylinder guide, gas cylinder sleeve and spring (fig. 56).

j. **Breechblock Locking Key.** Remove the locking wire, the breechblock locking plate screws with lock washers, the plate, (fig. 57).

39. ASSEMBLY AND REPLACEMENT.
a. Prior to assembly, all parts must be free of dirt, rust, and other extraneous matter. Metal parts in contact must be covered with a light film of lubricating oil. Assembly and replacement are in the reverse order of disassembly and removal. However, the following instructions pertaining to certain assembly operations should be noted:

(1) The breechblock lock must be assembled to the breechblock by collapsing the breechblock slides and, at the same time, exerting pressure against the lock until it is in the unlocked position. Hold the breechblock firmly in this position and push into the receiver as far as it will go so that the lock will not spring out of position (fig. 14).

(2) To assemble the driving spring and driving spring guide, first pry the sear down with a screwdriver and move the breechblock to the locked position. Insert the driving spring assembling tool into the driving spring guide plunger. Slip the driving spring over the tool and plunger. Insert the driving spring guide into the spring so that the tool telescopes

TM 9-227

DISASSEMBLY AND ASSEMBLY

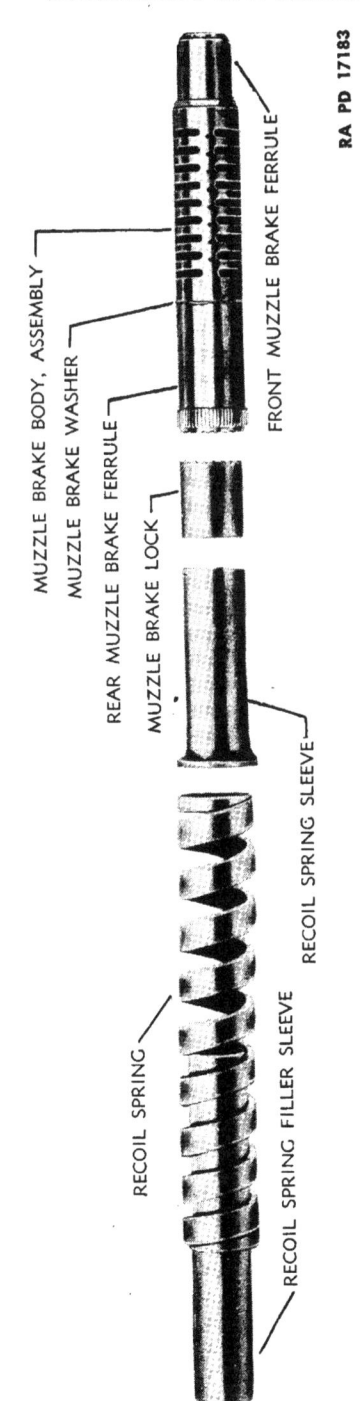

Figure 54 — Muzzle Brake and Recoil Spring Group

TM 9-227
39

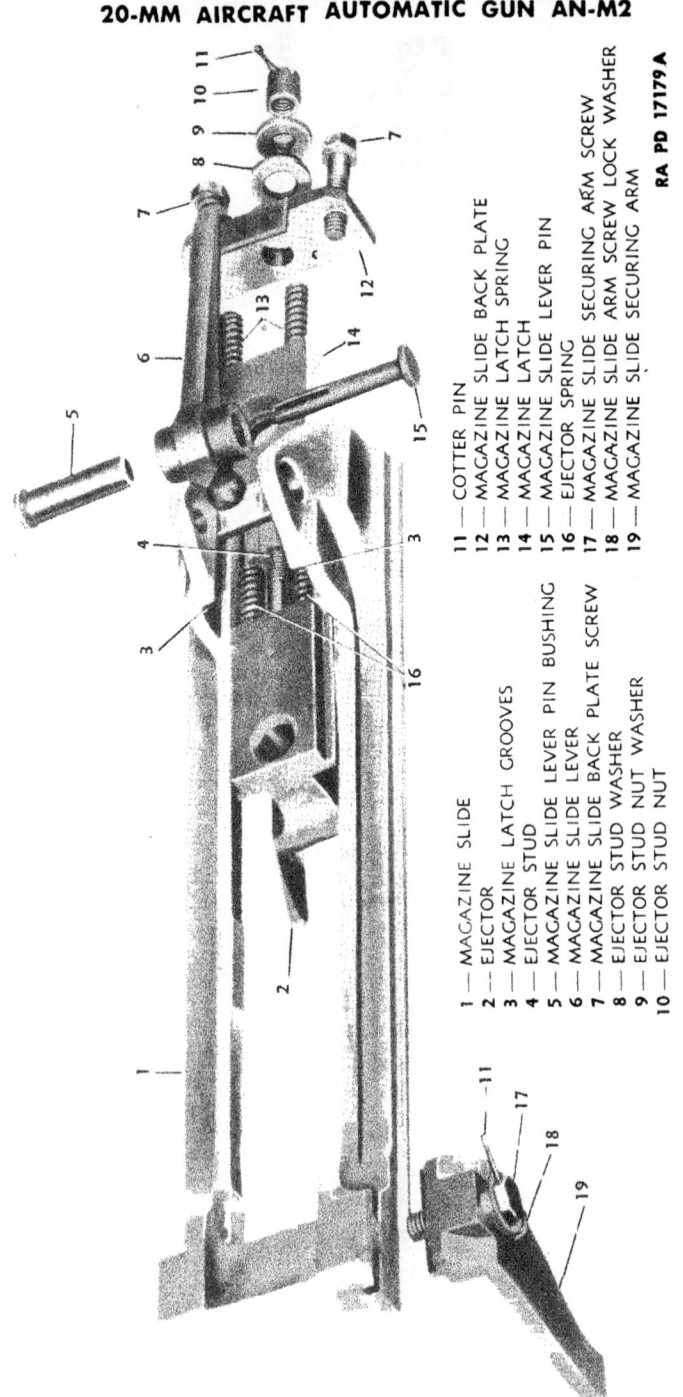

20-MM AUTOMATIC GUN M1 AND
20-MM AIRCRAFT AUTOMATIC GUN AN-M2

RA PD 17179 A

1 — MAGAZINE SLIDE
2 — EJECTOR
3 — MAGAZINE LATCH GROOVES
4 — EJECTOR STUD
5 — MAGAZINE SLIDE LEVER PIN BUSHING
6 — MAGAZINE SLIDE LEVER
7 — MAGAZINE SLIDE BACK PLATE SCREW
8 — EJECTOR STUD WASHER
9 — EJECTOR STUD NUT WASHER
10 — EJECTOR STUD NUT
11 — COTTER PIN
12 — MAGAZINE SLIDE BACK PLATE
13 — MAGAZINE LATCH SPRING
14 — MAGAZINE LATCH
15 — MAGAZINE SLIDE LEVER PIN
16 — EJECTOR SPRING
17 — MAGAZINE SLIDE SECURING ARM SCREW
18 — MAGAZINE SLIDE ARM SCREW LOCK WASHER
19 — MAGAZINE SLIDE SECURING ARM

Figure 55—Magazine Slide Group

DISASSEMBLY AND ASSEMBLY

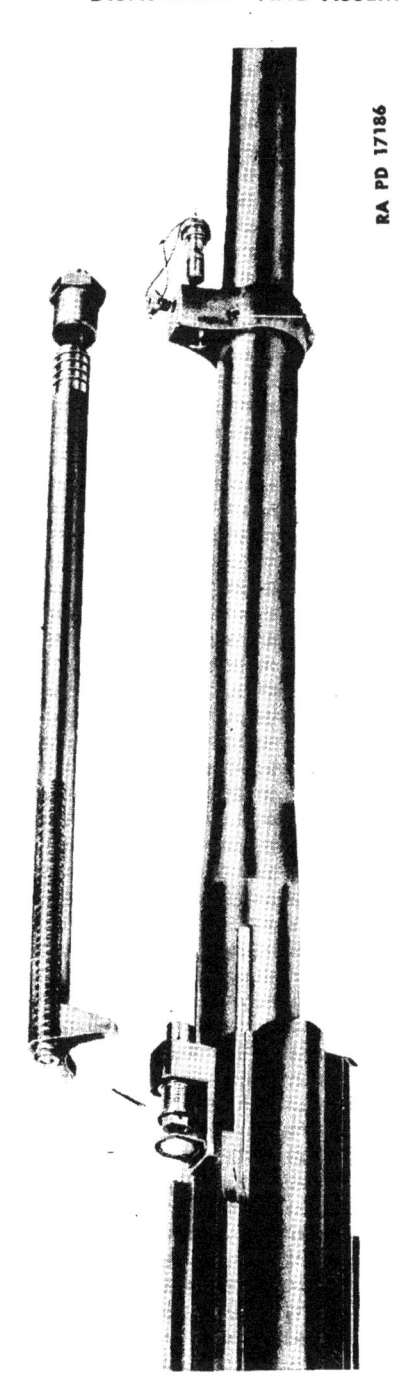

Figure 56 — Gas Cylinder and Sleeve Group

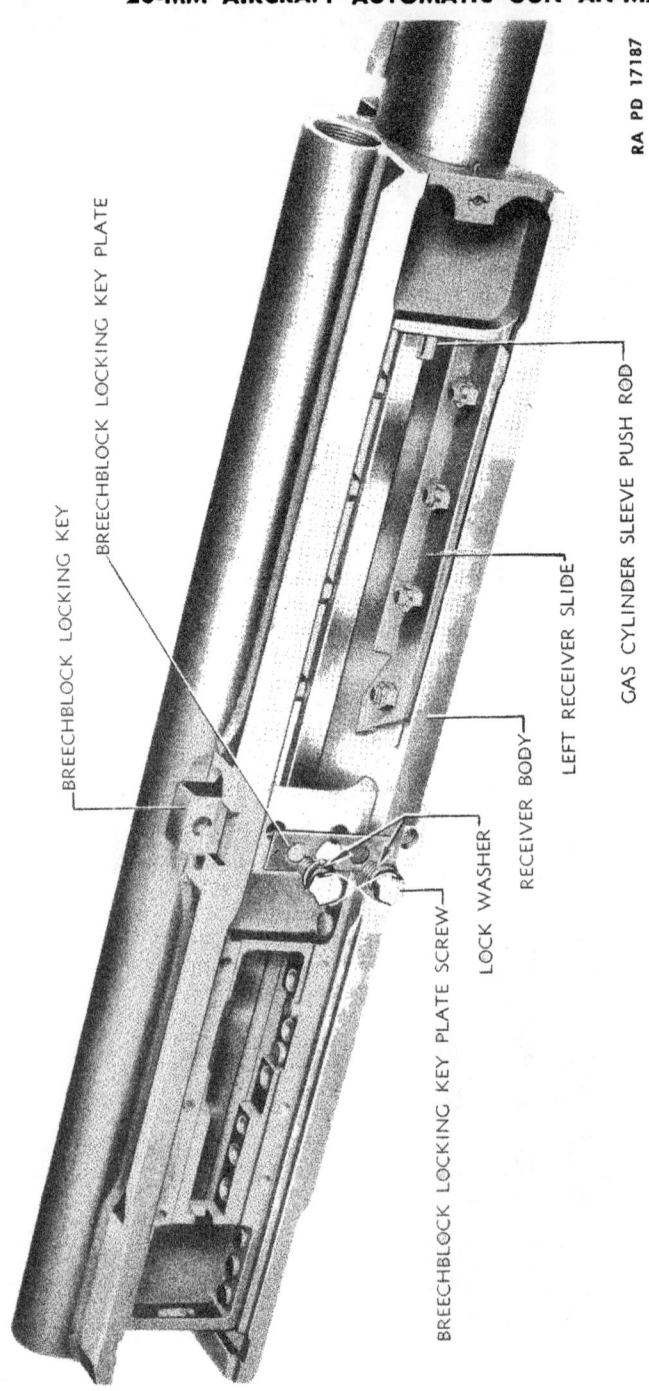

20-MM AUTOMATIC GUN M1 AND 20-MM AIRCRAFT AUTOMATIC GUN AN-M2

Figure 57 — Receiver Assembly

DISASSEMBLY AND ASSEMBLY

the guide tube. Insert the entire group into the receiver (rear buffer in place) (fig. 58), so that the head of the plunger rests against the back of the firing pin. Compress the driving spring and tighten the guide securely, using the rear buffer wrench (fig. 39). Remove the driving spring assembling tool and stake retainer into guide head.

Figure 58—Inserting the Driving Spring Group

(3) When replacing the muzzle brake, screw it on as far as it will go, making certain that all sleeves are properly assembled. Coat the threads and serrations with COMPOUND, antiseize, white lead base, or a castor oil flake graphite mixture before assembling.

(4) In assembling and replacing the sear block group, proceed as follows: Insert the sear buffer springs in their recesses in the sear block. Then, replace the plungers with their hollow ends against the springs. Place the unit on the sear block assembling tool with the flanged side up and with the radial bearing surface against the jaw of the tool. Compress the springs until the sear buffer spring retaining tool can be inserted to engage the grooves of the plungers (fig. 52). Do not remove the retaining tool until the group has been replaced in the gun.

b. **Sear Cover Plate Assembly.** Be sure to use two lock washers with the two screws nearest the trigger housing. Carefully lift the block from the tool and attach the sear. Assemble safety trigger pin with safety lever deep notch toward the outside of the sear cover plate. Otherwise the safety lever pin will be locked in the plate by the ball bearing.

40. DISASSEMBLY OF 60-ROUND MAGAZINE M1.

a. Remove all rounds from the magazine. Place the magazine in the magazine holder with the mouth up and the nut on the lower tie rod in the locating hole in the holder. If no magazine holder is available, use any other suitable retaining device.

b. Remove the cotter pin from the tensioning tube pin. Place a bar in the end of the tensioning tube and turn it to take the load off the tension tube pin. Remove the tube pin and carefully release the spring.

20-MM AUTOMATIC GUN M1 AND
20-MM AIRCRAFT AUTOMATIC GUN AN-M2

c. Unscrew the seven fixing screws and remove the front plate disk. Remove the spring casing with the spring and tensioning tube. Disengage the tensioning tube from the spring.

d. Remove the pin and collar from the rear of the feed arm axis tube. Turn the feed arm axis tube through a right angle so that the follower clears the hole in the front plate and remove the tube with feed arm and follower (fig. 59).

41. ASSEMBLY OF 60-ROUND MAGAZINE M1.

a. To assemble the magazine, proceed in the reverse order of disassembly. To apply initial tension to the magazine, proceed as in paragraph 23 a.

42. DISASSEMBLY OF 20-MM FEED MECHANISM M1 (FOR RIGHT-HAND FEED).

a. Remove the front and rear cover screws at the ends of the mouth. Remove the tie rod nuts and lock washers. Pull off the front and rear covers (fig. 60). The case may snap outward when the covers are released. The mouth will drop off, but the shaft assembly will remain in the case (figs. 61 and 62).

b. Remove the retainer spring and feed lever spring pin which fastens the springs through the case near the rear of the lower right side. Remove the shaft assembly from the case (fig. 63).

c. If the hub, ratchet actuating segment, and driving spring tensioning ratchet have not been removed from the cover on the end of the shaft, push them out rearward.

d. Hold the rack in place, unscrew the rack retaining screw, and gradually release the rack assembly, rack spring, and rack spring guide.

e. Remove the ratchet spring retaining plug from the upper side of the cover and withdraw the tensioning ratchet pawl and the tensioning ratchet pawl spring. Remove the tensioning ratchet pawl retaining screw.

f. Drift out the hub retaining pin to release the hub. Next, withdraw the ratchet actuating segment, thrust spring, and driving spring case cover from the shaft.

g. Drift out the collar retaining pin to release the collar at the rear of the shaft. Withdraw the rear feed lever, rear feed sprocket, front feed lever, center feed sprocket, and ejector assemblies. Drift out the front feed sprocket bushing pin to release the shaft key if it is necessary to remove the front feed sprocket assembly. To release the link chute cover, withdraw the tie rod.

DISASSEMBLY AND ASSEMBLY

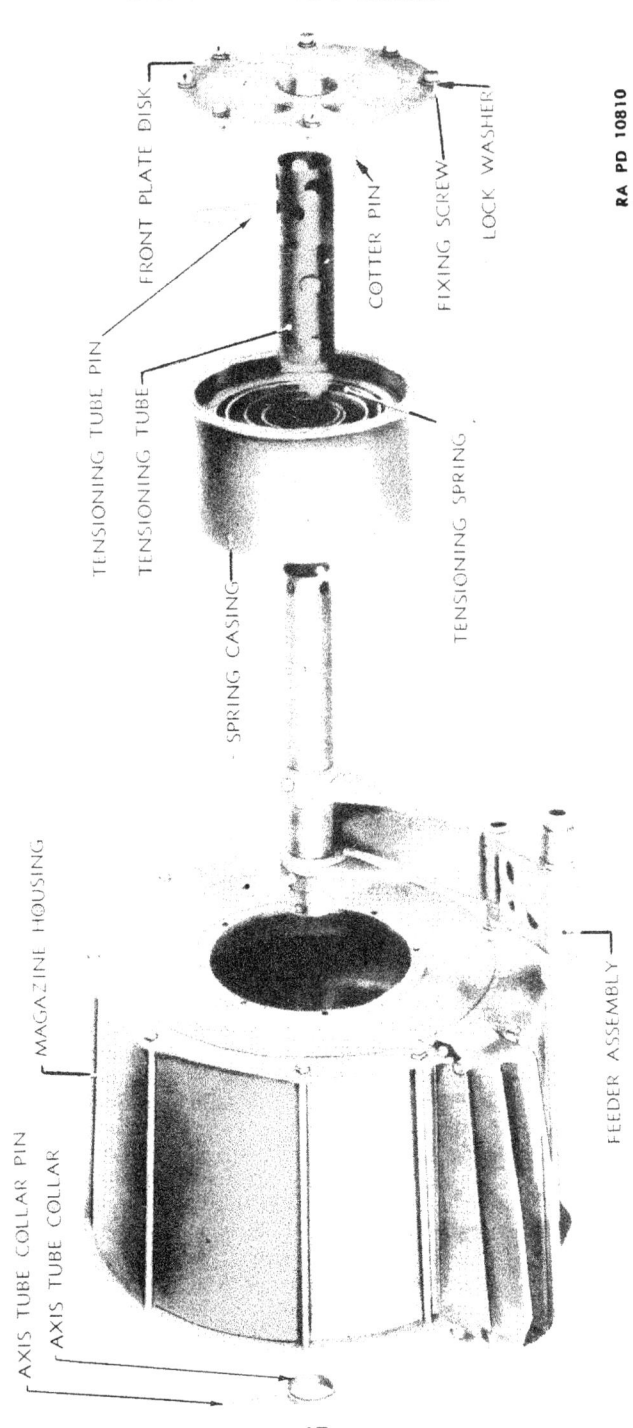

TM 9-227
42

Figure 59 — 20-mm 60-Round Magazine M1

TM 9-227
42

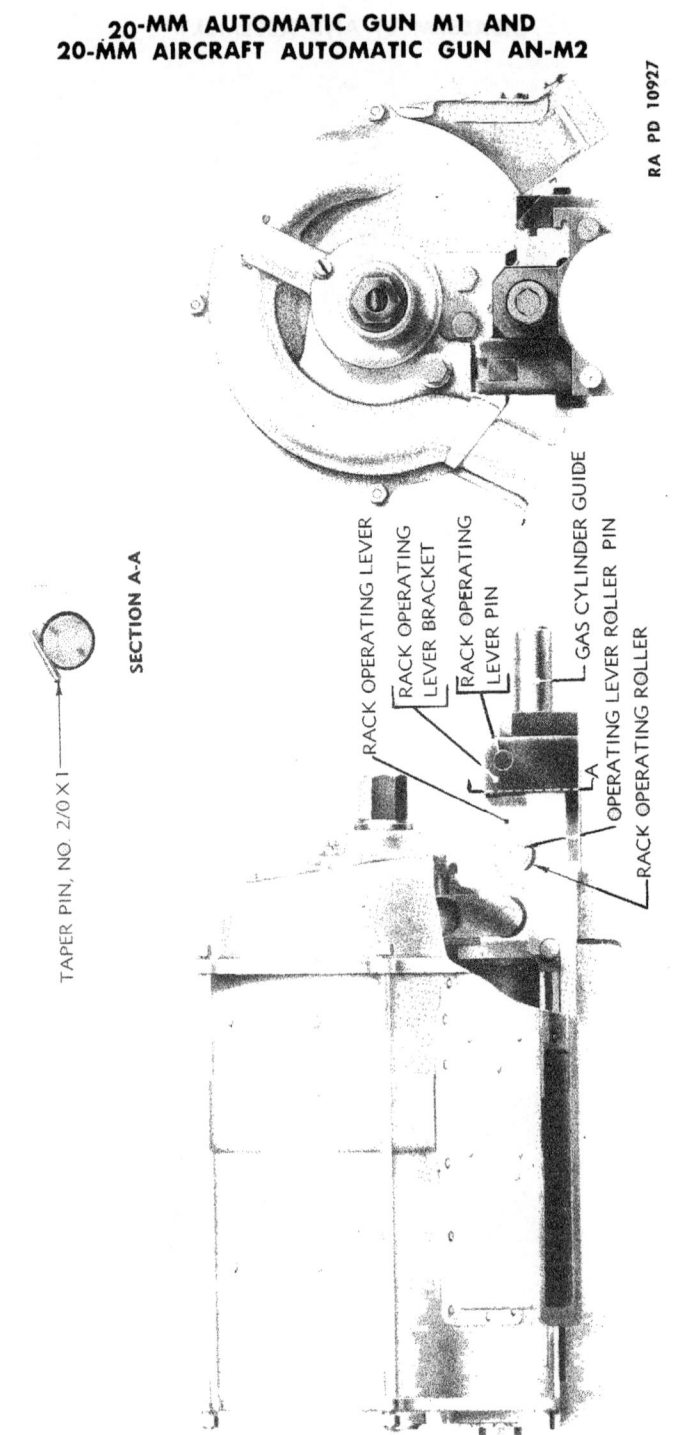

**20-MM AUTOMATIC GUN M1 AND
20-MM AIRCRAFT AUTOMATIC GUN AN-M2**

Figure 60 — 20-mm Feed Mechanism M1 and Rack Operating Lever Bracket Assembly

DISASSEMBLY AND ASSEMBLY

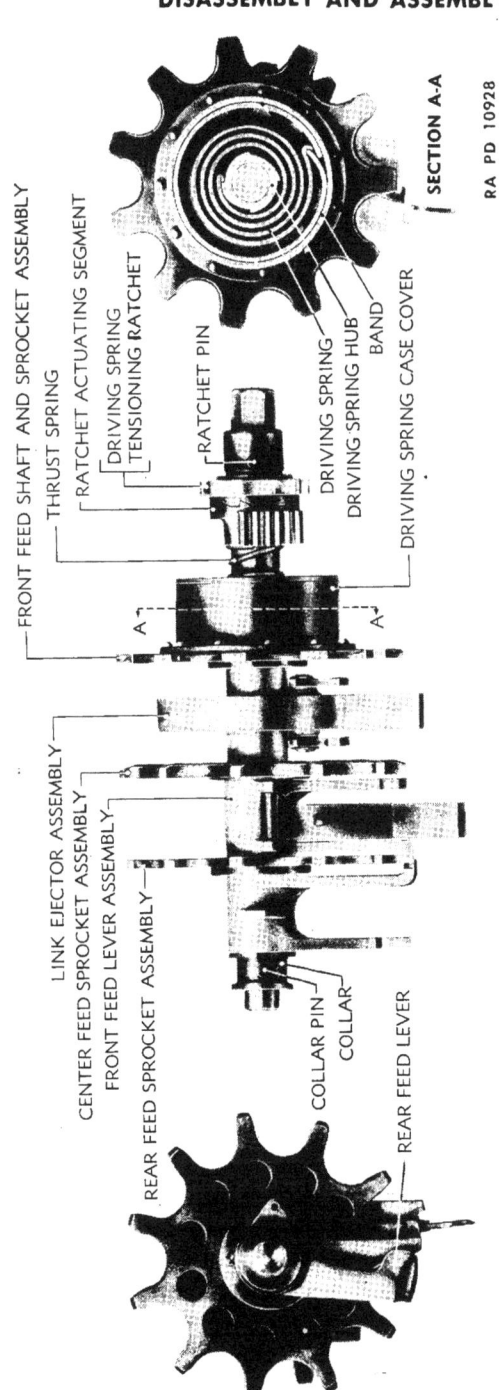

Figure 61—Feed Mechanism With Cover Removed

TM 9-227
42

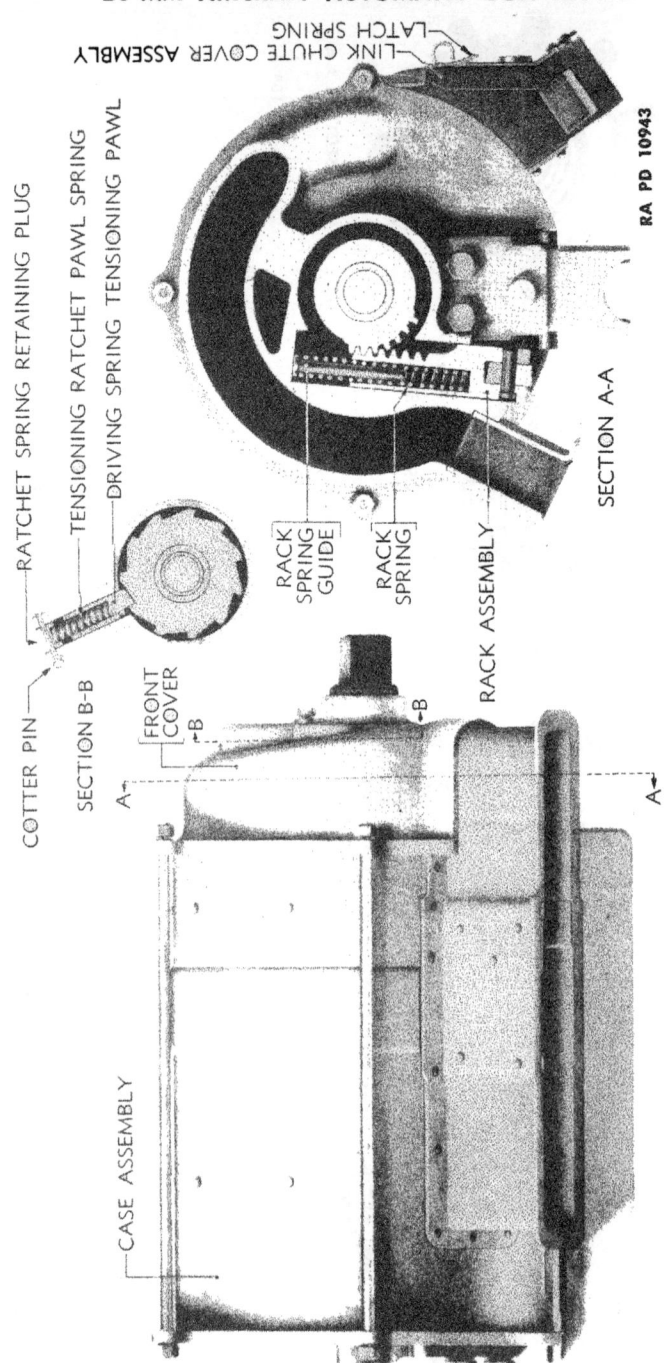

**20-MM AUTOMATIC GUN M1 AND
20-MM AIRCRAFT AUTOMATIC GUN AN-M2**

Figure 62—Front Cover Sectional Views

DISASSEMBLY AND ASSEMBLY

43. ASSEMBLY OF 20-MM FEED MECHANISM M1 (FOR RIGHT-HAND FEED) (figs. 61 and 62).

a. If the front feed sprocket assembly has been removed, slip it on the shaft and position it by replacing the shaft key. Drive the front feed sprocket bushing pin through the bushing and shaft. Next, place the following assemblies on the shaft in the order listed: ejector assembly, with ejector to the left; center feed sprocket assembly, with bushing to the rear; front feed lever assembly, with last round retainer to the right; rear feed sprocket assembly, with bushing to the rear; and rear feed lever. Fasten the entire group by driving the collar retaining pin through the collar and shaft.

b. Place the driving spring case cover on the shaft and follow with the thrust spring, ratchet actuating segment, and tensioning ratchet. Drive the hub retaining pin through the tensioning ratchet and the hub.

c. Replace the rack spring in its recess in the rack assembly and insert the rack spring guide into the rack spring. Place this unit in the recess provided in the front cover. Aline the teeth of the rack front to rear, and fasten it with the rack retaining screw.

d. Insert the hexagonal end of the tensioning ratchet through the central bore of the front cover from the rear. Engage the teeth of the rack and segment. The first tooth on the rack must engage between the first and second teeth of the segment.

e. Insert the tensioning ratchet pawl into the bore near the top of the cover. Aline the pawl and insert the tensioning ratchet pawl from the front of the cover. Insert the tensioning ratchet pawl spring on top of the pawl and close the bore with the pawl spring retaining plug.

f. Attach the last round retainer spring to the stud on the last round retainer. Attach the rear feed lever spring to the stud on the rear feed lever.

g. Place the shaft in the case assembly. The belt guide should project to the right when assembled. Slip the ends of the springs through the small slots provided in the case below the belt guide and fasten them with the last round retainer spring pin and rear feed lever spring pin.

h. Place the front cover on the forward end of the shaft and the rear cover assembly on the rear end of the shaft. Compress the case until the tie rods can be inserted through the drilled ears of the covers. One tie rod forms the hinge pin for the link chute cover. It may be easier to attach this rod before assembling the covers.

i. Slip the mouth into position with the slanted end toward the rear. Compress the case until the edges enter the grooves in the flanges of the

TM 9-227
43

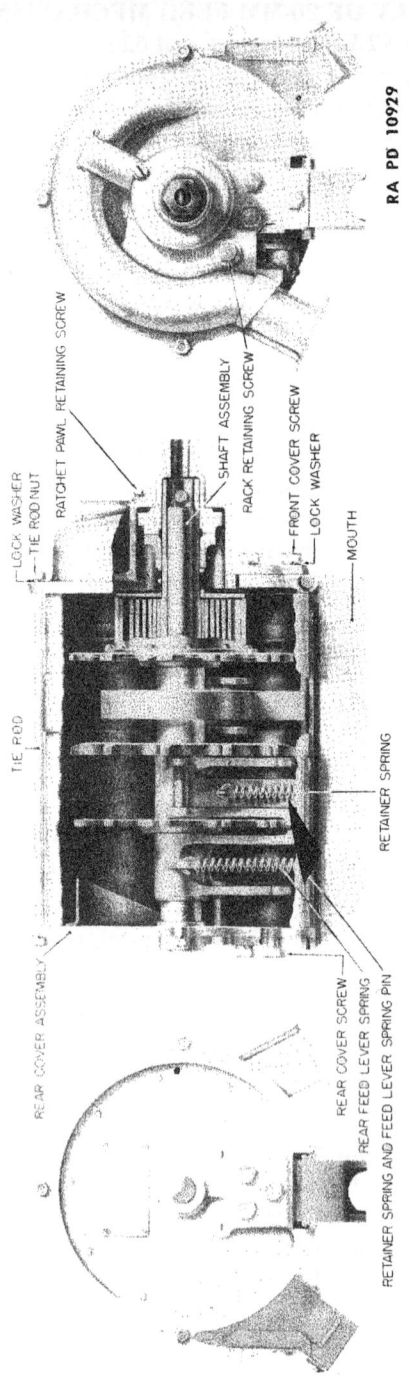

**20-MM AUTOMATIC GUN M1 AND
20-MM AIRCRAFT AUTOMATIC GUN AN-M2**

Figure 63 — Sectional View of Feed Mechanism

DISASSEMBLY AND ASSEMBLY

mouth and in the front and rear covers. Fasten the tie rods with lock washers and nuts; fasten the mouth to the covers with lock washers and the front and rear cover screws. Fasten the tie rod nuts gradually and uniformly, otherwise excessive friction will result and cause reduced recoil of the gun with attendant loss of spring tension.

NOTE: The last round retainer should extend downward within the mouth. The notch in the end of the link ejector should engage the right edge of the mouth. The rear feed lever should extend downward to the right of the rear feed lever stop.

j. After the mechanism has been assembled, press the rack two or three times with the thumb and see whether the feed sprockets rotate freely, moving one complete tooth for each full movement of the rack. If not, adjust the tie rod nuts and repeat the test until the feed sprockets rotate freely.

44. DISASSEMBLY AND ASSEMBLY OF 20-MM FEED MECHANISM M1 (FOR LEFT-HAND FEED).

a. Disassembly and assembly of the 20-mm feed mechanism for left-hand feed is the same as for right-hand feed except that the left-hand parts are in the reverse positions of the right-hand parts.

20-MM AUTOMATIC GUN M1 AND 20-MM AIRCRAFT AUTOMATIC GUN AN-M2

Section VII

INSPECTION

	Paragraph
General	45
Inspection of feeder	46
Driving spring guide group	47
Rear buffer group	48
Breechblock group	49
Sear cover plate group	50
Sear block group	51
Magazine slide group	52
Gas cylinder and sleeve group	53
Muzzle brake group	54
Recoil spring and mounting sleeve group	55
Receiver group	56
Tube	57

45. GENERAL.

a. The inspection procedure outlined below should be carefully carried out before and after each session of firing.

b. Clean and oil in accordance with instructions in section V before proceeding with inspection.

c. Cock the gun, noting smoothness of action.

d. Look for foreign matter in receiver.

e. Note lubrication, which *must* not be excessive.

46. INSPECTION OF FEEDER.

a. Examine exterior of feeder (magazine or feed mechanism) for loose or broken parts. Test whether the tie rods and nuts are tight.

b. If the case or covers are dented or damaged, and if the lips on the mouth are bent, the feeder is unserviceable.

c. Remove any burs from the mouth, from the pins at the front of the mouth and from the latch plate at the rear of the mouth.

d. If the 60-round magazine has been disassembled, apply initial tension to the spring (par. 23 a).

e. If the feed mechanism has been disassembled, test whether the feed sprockets rotate freely. Then, raise the link chute cover and see if spring is broken. Check whether rack roller rotates freely.

INSPECTION

f. Attach the feeder to the magazine slide and test for security of attachment.

47. DRIVING SPRING GUIDE GROUP.

a. Remove guide and note condition of threads on guide head.

b. Check guide tube for looseness in head, and also for deformations. Note condition and tightness of locking screw in guide head. Locking screw should be staked.

c. Examine driving spring for any sharp kinks or offset of coils which might cause binding or excess friction. Check free length of spring (26.5 in.). Replace if free length is less than 22.5 inches.

d. Check driving spring guide plunger for straightness. Note general condition of head, and look for cracks and indications of fracture just in rear of head. Test to see that plunger moves freely in and out of driving spring guide tube.

48. REAR BUFFER GROUP.

a. Retract and release rear buffer lock plunger, and note spring tension and fit of lock. Check free length of spring (0.82 in.). Replace if free length is less than 0.81 inches.

b. Remove rear buffer group and note any binding, looseness, or rough and bruised surfaces on dovetail connection. NOTE: If the rear buffer is provided with a retainer assembly (fig. 46), see that retainer is securely crimped to rear buffer housing. Note condition of retainer; see if pin on face of retainer is broken, bent, or missing.

49. BREECHBLOCK GROUP.

a. Note freedom of breechblock assembly in receiver. Remove breechblock and examine all parts, while disassembled, for burs and rough or rusty surface.

b. Test all springs for tension, kinks, and distortions.

c. Carefully check freedom of movement and general condition of extractor. Carefully examine the extracting surfaces. Check free length of spring (0.61 in.). Replace if free length is less than 0.60 inches.

d. Examine front face of bolt for erosion and wear, and note condition of firing pin hole. Measure diameter of firing pin hole, which should be 0.138 plus 0.004 inches.

e. Remove breechblock slides and examine for burs or rough surfaces on cams in rear and on outer surfaces near front of inertia block recess. Note condition of breechblock slide key. See if taper pin is in place in

20-MM AUTOMATIC GUN M1 AND
20-MM AIRCRAFT AUTOMATIC GUN AN-M2

right-hand slide and if key is tight in slide. Check free length of springs (2.80 in.). Replace if free length is less than 2.77 inches.

f. Remove firing pin and check freedom of movement in bolt. Examine point of pin for deformation, small cracks, or pitting.

g. Examine breechblock lock carefully for condition of cams on both sides and for wear or roughness on locking surface. Check underside of sear surface for wear.

h. Remove inertia blocks and note their general condition. Check movement of inertia block plungers and breechblock slide guides. Test tension of springs. Check free length of slide spring (2.80 in.) and of plunger spring (0.70 in.). Replace if free length is less than 0.69 inches.

50. SEAR COVER PLATE GROUP.

a. Remove group and note fit of cover to receiver. Check sear spring housing for looseness, and examine interior of housing for foreign matter or rough edges on the opening.

b. Test tension of sear spring. Check free length of spring (1.085 in.). Replace if free length is less than 1.07 inches.

c. Remove safety lever and check cam on its forward end.

d. Remove safety trigger pin and examine for burs or rough surfaces around the point where it engages bowden connection shaft, and also around the seats for the ball. Note action of the positioning ball in relation to tension and locking action as the pin is moved in and out. Check free length of ball spring (0.35 in.). Replace if less than 0.34 inches.

e. Remove the two nuts from bowden connection shaft and examine the parts for wear, rust, and broken or cracked bushings.

f. Examine bowden connection shaft spring for wear and free length (0.750 in.). Replace if free length is less than 0.74 inches.

g. Remove bowden connection shaft and check free movement in cover. Examine recess for the safety lever pin and cam surfaces on upper end for wear or burs.

51. SEAR BLOCK GROUP.

a. Remove sear group. Disconnect the sear. Tighten sear block in the assembling tool or in a vise just enough to remove the retaining rod (fig. 53), and then gradually release the pressure on the plungers. Check free length of sear spring (1.08 in.). Replace if less than 1.06 inches.

b. Examine sear surface for wear or roughness. Check plunger for burs or rough surfaces around the disassembling recesses. Check free

INSPECTION

length of buffer springs (1.89 in.). Replace if free length is less than 1.87 inches.

c. Note position of sear buffer blocks in receiver. The steel block should be in position adjacent to plungers. Note condition of blocks. Flat surface of fiber block should be adjacent to steel block.

52. MAGAZINE SLIDE GROUP.

a. Note condition of the magazine slide securing arm.

b. Remove ejector stud nut and then the ejector with its springs. Examine ejector and check freedom of movement in slide, without the springs. Note binding or excessive wear.

c. Check free length of ejector springs, normally 1.89 inch.

d. Examine ejector stud washer for deformations or indications of fracture.

e. Remove magazine slide backplate and disassemble magazine latch. Examine latch carefully for burs and excessive wear on underside.

f. Test tension of latch springs. Check free length of spring (2.83 in.).

g. Check freedom of movement of slide.

Table of Spring Dimensions

NAME	Mean Diameter Free	Wire Diameter	Free Height	Minimum Allowable Free Height
Driving spring	0.60 in.	0.100 in.	25.5 in.	23.5 in.
Inertia block plunger spring	0.216 in.	0.095 in.	.70 in.	.63 in.
Sear buffer spring	0.362 in.	0.266 in.	1.89 in.	1.70 in.
Trigger locking pin ball spring	0.165 in.	0.023 in.	.355 in.	.320 in.
Sear spring	0.262 in.	0.051 in.	1.085 in.	.977 in.
Rear buffer lock spring	0.315 in.	0.091 in.	.82 in.	.74 in.
Magazine latch spring	0.242 in.	0.064 in.	2.83 in.	2.55 in.
Gas cylinder sleeve spring	0.549 in.	0.135 in.	7.48 in.	6.73 in.
Extractor spring	0.302 in.	0.063 in.	.61 in.	.55 in.
Ejector spring	0.220 in.	0.072 in.	1.89 in.	1.70 in.
Breechblock slide spring	0.276 in.	0.054 in.	2.80 in.	2.52 in.
Bowden connection shaft spring	0.542 in.	0.056 in.	.75 in.	.68 in.
Recoil spring	—	—	12.625 in.	11.463 in.
Rear buffer spring	—	—	4.312 in.	3.881 in.

20-MM AUTOMATIC GUN M1 AND
20-MM AIRCRAFT AUTOMATIC GUN AN-M2

53. GAS CYLINDER AND SLEEVE GROUP.

a. Operate gas cylinder sleeve by hand and note any binding action.

b. Test tension of gas cylinder sleeve spring. Check free length (7.48 in.). Replace if free length is less than 7.40 inches.

c. Examine forward portion of gas cylinder guide for rust or rough surfaces. Check for carbon or rust on sleeve, yoke, piston, vent plug, and gas vent.

54. MUZZLE BRAKE GROUP.

a. Remove muzzle brake assembly (or thread protector) and examine for wear, rust, carbon, and broken or missing parts.

55. RECOIL SPRING AND MOUNTING SLEEVE GROUP.

a. Check free length of spring (12.60 in.). Replace if free length is less than 12.47 inches.

56. RECEIVER GROUP.

a. Remove push rods from the front and examine for burs or swedging of the ends.

b. Examine right and left receiver plates for tightness and check for burs or rough surfaces on locking cams.

c. Examine the breechblock locking key for wear and looseness in receiver.

d. Examine rear buffer guides in rear of receiver for cracks or signs of spreading or improper fitting of rear buffer.

57. TUBE.

a. Examine bore visually from both ends and note sharpness of lands, carbon deposits, powder fouling, rust, and coppering.

b. Examine chamber for rust, carbon, and pits.

Section VIII

AMMUNITION

	Paragraph
General	58
Nomenclature	59
Firing tables	60
Classification	61
Identification	62
Care, handling, and preservation	63
Authorized rounds	64
Preparation for firing	65
CARTRIDGE, S.A., H.E.I., Mk. I, w/FUZE, percussion, D.A., No. 253, Mk.I/A/, 20-mm auto. guns M1, AN-M2, and Hispano A	66
CARTRIDGE, S.A., A.P., M75, w/TRACER, 20-mm auto. guns M1, AN-M2, and Hispano /A/	67
CARTRIDGE, S.A., ball, Mk. I, 20-mm auto. guns M1, AN-M2, and Hispano /A/	68
Fuzes	69
FUZE, percussion, D.A., No. 253, Mk. I/A/	70
Packing and marking	71
Field report of accidents	72

58. GENERAL.

a. The ammunition for these guns is issued in the form of fuzed complete rounds of fixed ammunition. The term "fixed," used in connection with ammunition, signifies that the propelling charge is not adjustable and that the round is loaded into the gun as a unit. The propelling charge is assembled loosely in the cartridge case which is crimped rigidly to the projectile. A complete round comprises all the ammunition components used to fire a weapon once. After firing, the cartridge case is extracted and ejected, then the next round is loaded into the gun, all automatically.

59. NOMENCLATURE.

a. Standard nomenclature is used herein in all references to specific items of issue. Its use for all purposes of record is mandatory.

60. FIRING TABLES.

a. Data on applicable firing tables and trajectory charts are not available.

61. CLASSIFICATION.

a. Dependent upon the type of projectile, the ammunition is classi-

20-MM AUTOMATIC GUN M1 AND
20-MM AIRCRAFT AUTOMATIC GUN AN-M2

fied as high explosive incendiary, armor-piercing, or ball. The high explosive incendiary projectile contains both a high explosive and an incendiary filler. The armor-piercing projectile is a solid shot, containing a tracer element for observation of fire, that is, for showing the gunner the path of the projectile in flight. The ball projectile is inert and is provided for use against personnel and light materiel targets.

62. IDENTIFICATION.

a. General. Ammunition, including components, is completely identified by means of painting and marking (including ammunition lot number). Other essential information may be obtained from the marking. See figures 64, 65, and 66 and the paragraphs below.

b. Mark or Model. To identify a particular design, a model designation is assigned at the time the design is classified as an adopted type. This model designation becomes an essential part of the standard nomenclature of the item and is included in the marking of the item. The model designation consists of the letter "M" followed by an arabic numeral. Modifications are indicated by adding the letter "A" and the appropriate arabic numeral. Thus, "M1A1" indicates the first modification of an item for which the original model designation was "M1." An exception exists in the case of some models of 20-mm ammunition which are designated "Mark," abbreviated "Mk.," followed by a Roman numeral.

c. Ammunition Lot Number.

(1) When ammunition is manufactured, an ammunition lot number, which becomes an essential part of the marking, is assigned in accordance with pertinent specifications. This lot number is stamped or marked on every complete round and on all packing containers. It is required for all purposes of record, including reports on condition, functioning, and accidents, in which the ammunition is involved. To provide for the most uniform functioning, all of the rounds of any one lot of fixed ammunition consist of:

(a) Projectiles of one lot number.

(b) Fuzes of one lot number.

(c) Primers of one lot number.

(d) Propellent powder of one lot number.

(2) Therefore, to obtain the greatest accuracy in any firing, successive rounds should be from the same ammunition lot whenever practicable.

d. Painting and Marking.

(1) PAINTING. Projectiles are painted to prevent rust and, by the color, to provide a ready means of identification as to type. The pro-

AMMUNITION

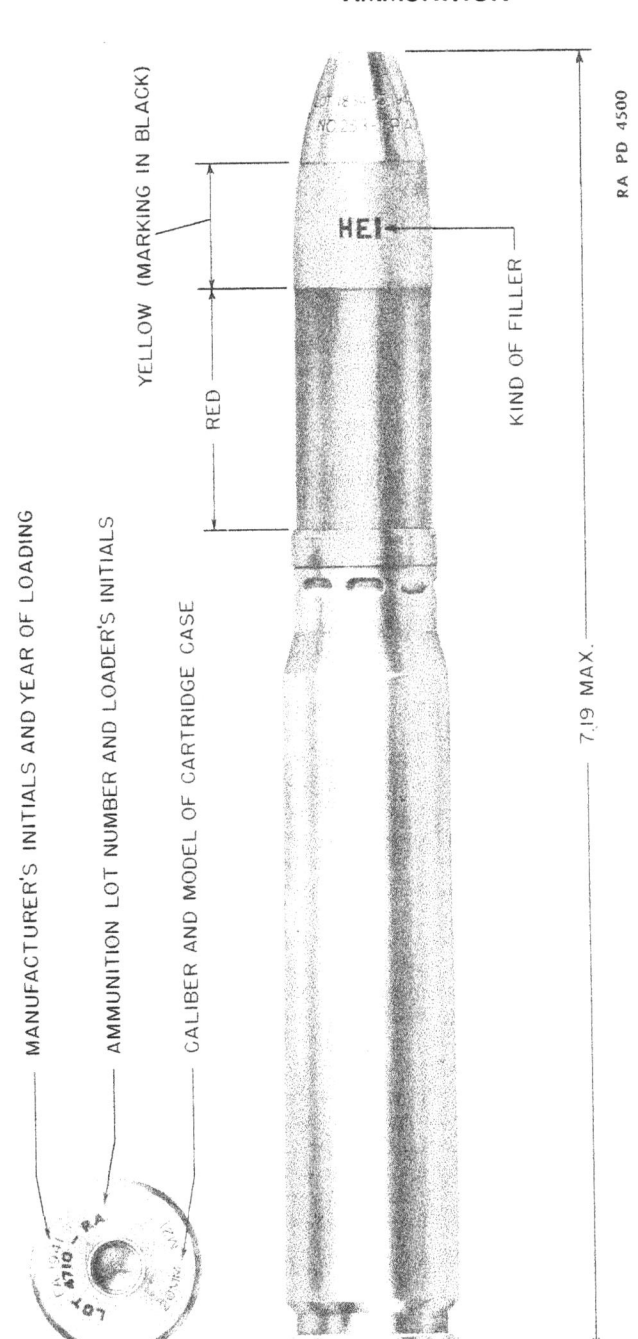

Figure 64—Cartridge, S.A., H.E.I., W/Fuze, Percussion, D.A., No. 253, Mk. I/A/, Mk. I, 20-mm Auto. Guns, M1, AN-M2, and Hispano /A/

20-MM AUTOMATIC GUN M1 AND
20-MM AIRCRAFT AUTOMATIC GUN AN-M2

jectiles of the ammunition described herein are painted as follows:

High explosive incendiary . yellow ogive, red body: marking in black.
Armor-piercing . . . . black: marking in white.
Ball (inert) . . . . black: marking in white.

NOTE: The above color scheme is not wholly in agreement with the basic color scheme described in TM 9-1900.

(2) MARKING. For purposes of identification, the following is marked or stamped on the components of each round of fixed ammunition described herein:

(a) On the Projectile (Stenciled):
1. On the H.E.I. projectile:
    Kind of filler.
2. On the A.P. projectile:
    Caliber and type of weapon in which fired.
    Model of projectile.
    "WITH TRACER."

(b) On the Projectile (Stamped in the Metal):
1. On the H.E.I. and practice projectiles (on the body):
    Manufacturer's initials or symbol.
    Lot number of empty projectile.
    Month and year of manufacture.
2. On the A.P projectile (on the base end):
    Manufacturer's initials or symbol.
    Lot number of projectile.
    Year of manufacture.
    Caliber and designation of shot.

(c) On the Head of the Cartridge Case:
1. Stenciled:
    Ammunition lot number.
    Loader's initials.
2. Stamped in the metal:
    Designation and caliber of case.
    Manufacturer's initials or symbol.
    Year of manufacture, in full.

(d) On the Fuze (Stamped in the Metal):
   Model and designation of fuze.
   Manufacturer's initials or symbol.
   Loader's lot number.
   Year of loading.

AMMUNITION

63. CARE, HANDLING, AND PRESERVATION.

a. Complete rounds are packed to withstand conditions ordinarily encountered in the field. Ammunition for the 20-mm automatic guns is packed in cartons (10 per carton), which in turn are inclosed in metal-lined wooden boxes. Since explosives are adversely affected by moisture and high temperature, the following precautions should be observed:

(1) Do not break moisture-resistant seals until ammunition is to be used.

(2) Protect ammunition, particularly fuzes, from high temperatures, including the direct rays of the sun. More uniform firing is obtained if all the rounds are at the same temperature.

b. Handle ammunition with care at all times. The explosive elements in primers and fuzes are highly sensitive to shock and high temperature.

c. Do not attempt to disassemble any complete round or fuze.

d. The complete round should be free of foreign matter—sand, mud, grease, etc.—just before loading into the magazine or belt. If it gets wet or dirty, it should be wiped up at once.

e. Although the use of oil or grease on ammunition is generally prohibited, in the case of ammunition for these guns it is necessary to oil the cartridge case in order to prevent jamming. By means of a cloth wrung out of OIL, lubricating, preservative, light, spread a light film of oil evenly over the body of the cartridge case just prior to insertion of the round into the magazine or belt. Extreme care should be taken to prevent oil from getting on the primer or joint at the mouth of the cartridge case. If OIL, lubricating, preservative, light, is not available, use OIL, lubricating, for aircraft instruments and machine guns. Preferably, only one day's supply of ammunition should be lubricated at a time. Rounds oiled for firing and not fired the same day, should be wiped dry to prevent the accumulation of dust and grit, and the seepage of oil around the primer and mouth of the cartridge case. Such rounds will be used first in subsequent firing; they must be oiled again before use.

f. Duds must not be handled as they are extremely dangerous. Dispose of them in accordance with the instructions in TM 9-1900.

64. AUTHORIZED ROUNDS.

a. The ammunition authorized for use in these guns is shown in the following table. The M1, AN-M2 and Hispano A Guns all are chambered alike, hence fire the same ammunition. The nomenclature (standard nomenclature) completely identifies the round.

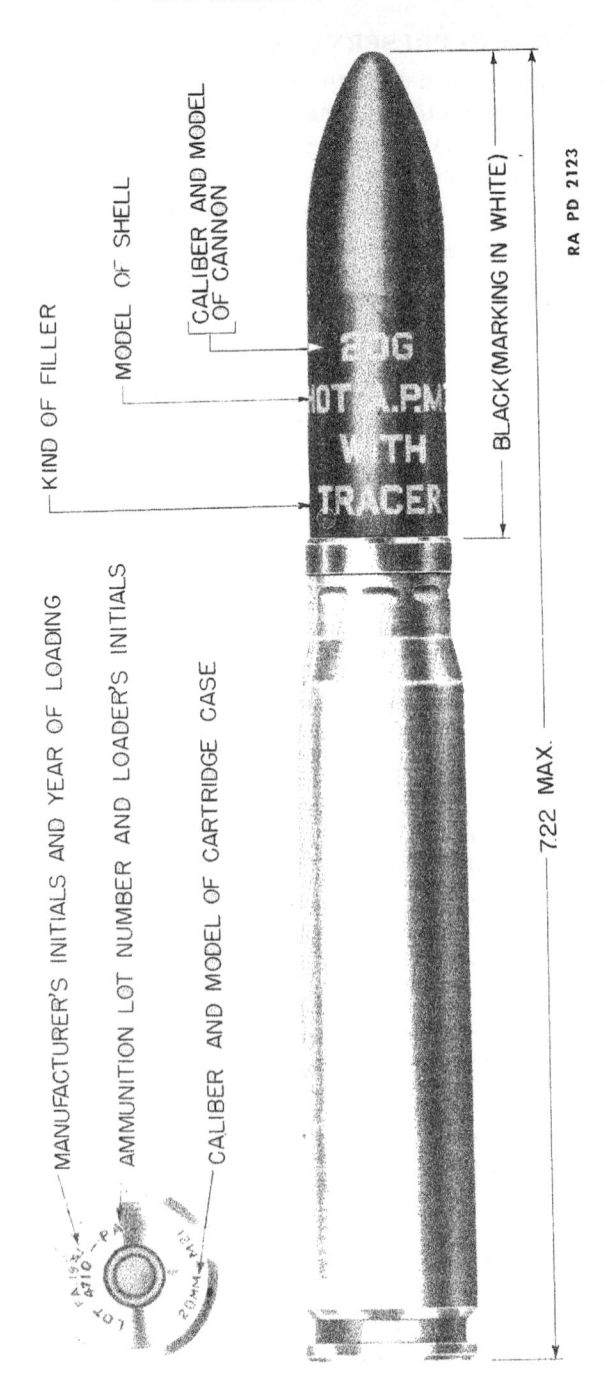

20-MM AUTOMATIC GUN M1 AND
20-MM AIRCRAFT AUTOMATIC GUN AN-M2

Figure 65 — Cartridge, S.A., Shot, A.P., M75 w/Tracer, 20-mm Auto. Guns, M1, AN-M2, and Hispano A

AMMUNITION

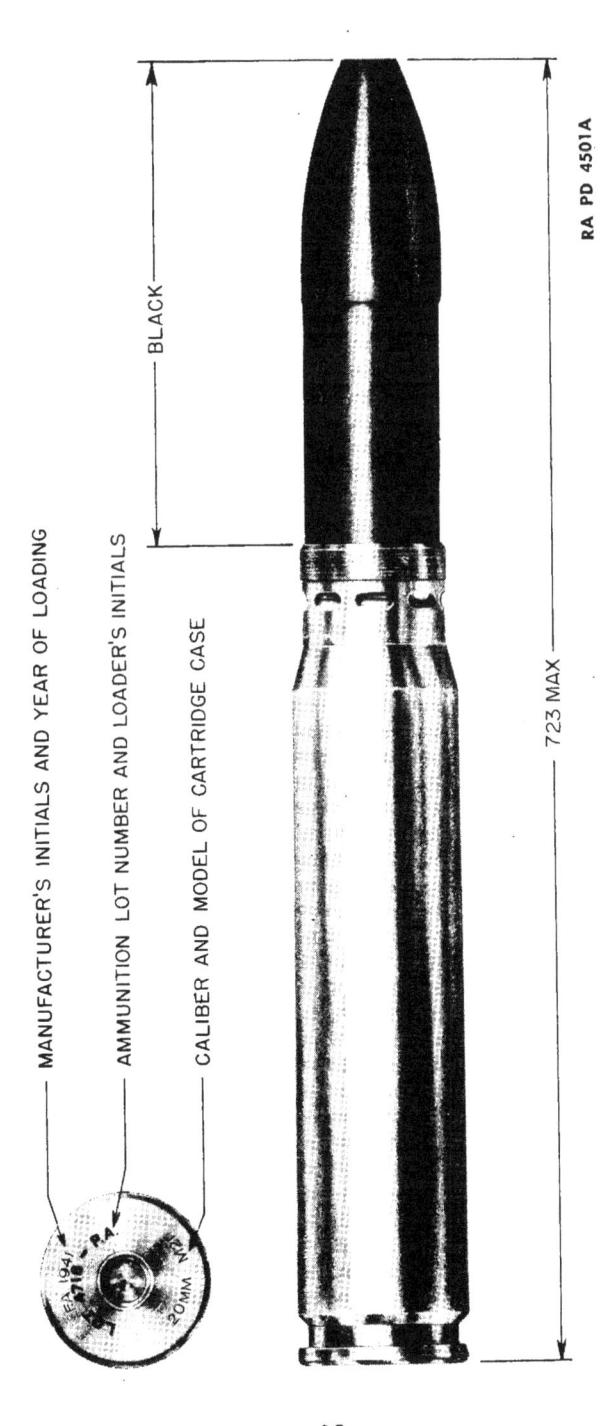

Figure 66—Cartridge, S.A., ball, Mk. I, 20-mm Auto. Guns, M1, AN-M2, and Hispano /A/

20-MM AUTOMATIC GUN M1 AND
20-MM AIRCRAFT AUTOMATIC GUN AN-M2

Table I

Ammunition for the Gun, Automatic, 20-mm, M1, AN-M2, and Hispano A

Nomenclature	Action of fuze	Approximate weight of projectile as fired
Service Ammunition		
CARTRIDGE, S.A., H.E.I., Mk. I, w/fuze, percussion, D.A., No. 253, Mk. I/A/, 20-mm auto. guns, M1, AN-M2 and Hispano /A/	Supersensitive	0.29 lb
CARTRIDGE, S.A., shot, A.P., M75, w/tracer, 20-mm auto. gun, M1, AN-M2 and Hispano /A/	None	0.37 lb
CARTRIDGE, S.A., Ball, Mk. I, 20-mm auto. guns, M1, AN-M2 and Hispano /A/	None	0.28 lb

H E.I — High explosive incendiary   A P — Armor-piercing

65. PREPARATION FOR FIRING.

a. As issued, the complete rounds are ready for firing after removal of packing; however, it is necessary to oil the rounds as described in paragraph 21 and to load the rounds into the feed mechanism or the magazine.

66. CARTRIDGE, S.A., H. E. I., MK. I, W/FUZE, PERCUSSION, D. A., NO. 253, MK. I/A/, 20-MM AUTO. GUNS M1, AN-M2, and Hispano /A/.

a. This complete round (fig. 64) is designed for use from aircraft against light materiel targets, functioning with both explosive and incendiary effect. After the shell penetrates the target, the high explosive filler is detonated, the shell is shattered, and the incendiary composition is ignited. The round consists of a primer and a propelling charge, contained in a brass cartridge case which is crimped rigidly to the projectile, and a fuze which is of the supersensitive type. The projectile contains a total of 0.03 pound of high explosive and incendiary fillers. The round is 7.19 inches long and weighs 0.57 pound. The propelling charge, weighing 0.07 pound, consists of loose flashless nonhygroscopic (FNH) smokeless powder contained in the cartridge case.

TM 9-227
67-68

AMMUNITION

67. CARTRIDGE, S.A., SHOT, A.P., M75, W/TRACER, 20-MM AUTO. GUNS M1, AN-M2, AND HISPANO /A/.

a. This complete round (fig. 65) is designed for use from aircraft against armored targets. It consists of a primer and a propelling charge contained in a brass cartridge case which is crimped rigidly to the projectile. The projectile is a solid steel shot and contains a red tracer composition in its base. The round is 7.22 inches long and weighs 0.64 pound. The propelling charge, weighing 0.07 pound, consists of loose flashless nonhygroscopic (FNH) smokeless powder contained in the cartridge case.

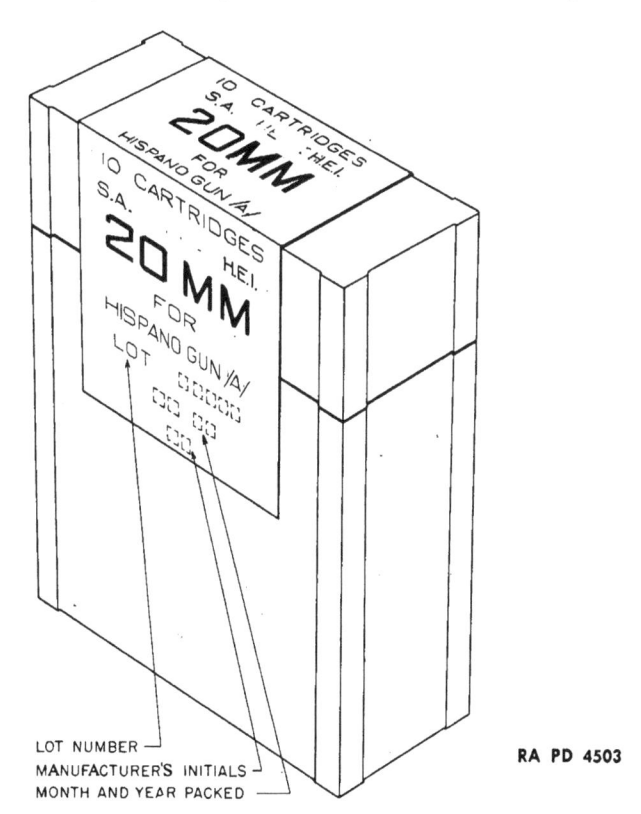

Figure 67 — Packing Carton for Ammunition for 20-mm Auto. Guns M1 and M2

68. CARTRIDGE, S.A., BALL, MK. I, 20-MM AUTO. GUNS M1 AN-M2, AND HISPANO /A/.

a. This complete round (fig. 66) is for service firing from aircraft against personnel and light materiel targets. It consists of a primer and propelling charge contained in a cartridge case which is crimped rigidly

TM 9-227
68–71

20-MM AUTOMATIC GUN M1 AND
20-MM AIRCRAFT AUTOMATIC GUN AN-M2

to the steel projectile. The projectile contains no explosive and has no fuze. It is similar in shape and ballistic properties to the point-fuzed high explosive incendiary projectile. The round is 7.23 inches long and weighs 0.56 pound. The propelling charge, weighing 0.07 pound, consists of loose flashless nonhygroscopic (FNH) smokeless powder contained in the cartridge case.

69. FUZES.

  a. A fuze is a mechanical device used with a projectile to explode it at the time and under the circumstances desired. A fuze designed to function upon impact with a target is classified as the impact type. Fuzes designed to function on impact with a light material target, such as an airplane wing, are further classified as supersensitive fuzes.

  CAUTION: Fuzes will not be disassembled. Any attempt to disassemble fuzes in the field is dangerous and is prohibited except under specific direction of the Chief of Ordnance.

70. FUZE, PERCUSSION, D. A., No. 253, MK. I/A/.

  a. This is a supersentive fuze of the impact type, designed to function just after penetration of light materiel targets. Like some fuzes used with small caliber ammunition, this fuze does not come within the definition of boresafe. It is used with 20-mm aircraft ammunition and is issued assembled to the high explosive incendiary projectile of the fixed complete round described in paragraph 66 and shown in figure 64.

71. PACKING AND MARKING.

  a. **Packing.** The ammunition for the GUN, automatic, 20-mm, M1 and AN-M2, is packed 10 rounds per fiber carton (fig. 67), 12 cartons (120 rounds) per sealed metal-lined packing box (fig. 68). The following data are considered suitable for estimating weight and volume requirements:

	Weight (pounds)	Volume (cubic feet)
Complete round, H.E.I., w/o packing material	0.57	....
Complete round, A.P., w/o packing material	0.64	....
Complete round, practice, w/o packing material	0.56	....
120 H.E.I. rounds in fiber cartons in metal-lined packing box	94.8	1.40
120 A.P. rounds in fiber cartons in metal-lined packing box	103.0	1.40
120 ball rounds in fiber cartons in metal-lined packing box	93.6	1.40

Over-all dimensions of packing box (in.):
   $18\frac{1}{8}$ x $13\frac{3}{16}$ x $10\frac{11}{32}$

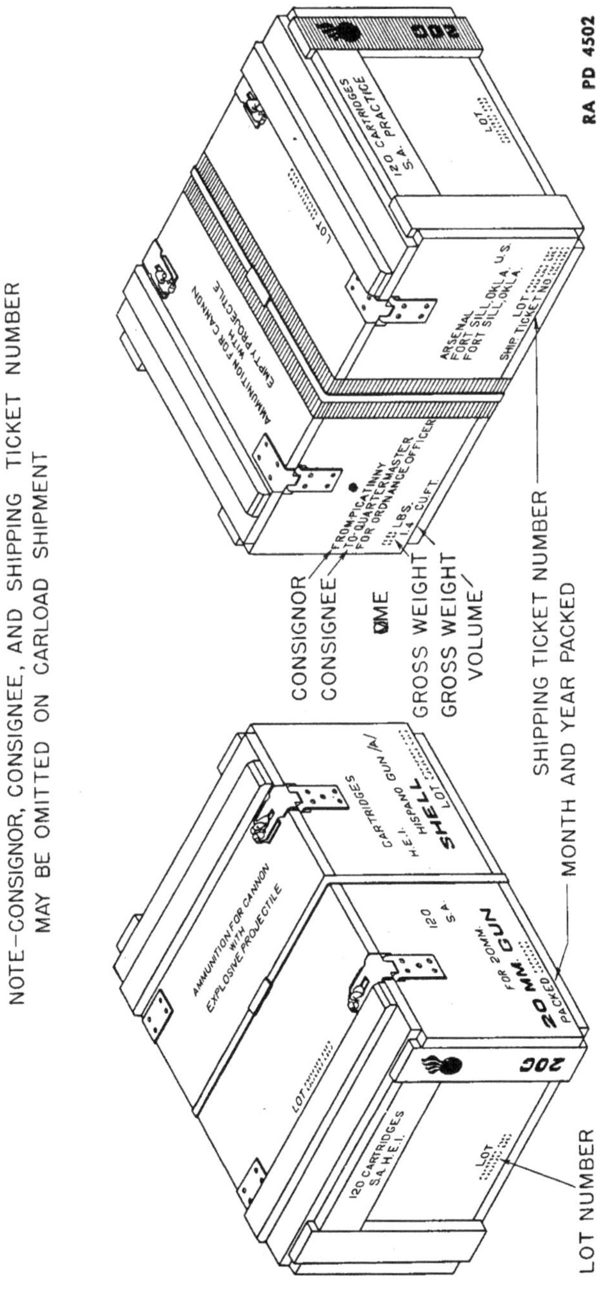

TM 9-227

AMMUNITION

Figure 68—Packing Box for Ammunition for 20-mm Auto. Guns M1 and AN-M2

20-MM AUTOMATIC GUN M1 AND
20-MM AIRCRAFT AUTOMATIC GUN AN-M2

b. Marking for Shipment.

(1) Packings for shipment are marked as follows (fig. 68):

(a) Name and address of consignee (or code marking).

(b) List and description of contents.

(c) Gross Weight in pounds, displacement in cubic feet.

(d) The number of the package.

(e) The letters "U.S." in several conspicuous places.

(f) Order number, contract number, or shipping number.

(g) Ordnance insignium and escutcheon.

(h) Name or designation of consignor preceded by the word "From."

(i) Lot number.

(j) Month and year packed.

(k) Inspector's stamp.

72. FIELD REPORT OF ACCIDENTS.

a. Any serious malfunctions of ammunition must be promptly reported to the ordnance officer under whose supervision the material is maintained or issued (par. 7, AR 45-30).

Section IX

ORGANIZATION SPARE PARTS AND ACCESSORIES

	Paragraph
Organization spare parts	73
Accessories	74

73. ORGANIZATION SPARE PARTS.

a. These are extra parts provided with the materiel for replacement of those parts which are most likely to become unserviceable through breakage or wear. Organization spare parts are for use by the using arm in making minor repairs. The sets of organization spare parts should be kept as complete as possible at all times and kept clean and oiled to prevent rust. The allowances of organization spare parts are prescribed in SNL A-47.

74. ACCESSORIES.

a. General. Gun accessories are those required for operating, disassembling, assembling, and for cleaning, care, and preservation. They also include covers, tool roll, etc. necessary for storage and protection when the equipment is not in use. Accessories should not be used for purposes other than as prescribed. Those accessories the names or general characteristics of which indicate their use are not described in detail here. Accessories embodying special features or having special uses are described in the following paragraphs:

(1) STAFF, CLEANING, M13, (20-MM). The rod consists of four metal sections threaded to each other and provided with a T-shaped handle at one end and a brush assembly at the other end. The brush assembly can be replaced with a plug end for use with a patch, or a loop end for a flannelette or other cleaning rag.

(2) TOOL, ASSEMBLING, DRIVING SPRING. This is a steel rod with a split stud at one end. The studded end is inserted through the driving spring guide tube into the recess in the outer end of the driving spring guide plunger to aid in removing and replacing the plunger in the bolt.

(3) TOOL, ASSEMBLING, SEAR BLOCK. This vise-like tool is used for compressing and releasing the sear buffer springs in disassembly and assembly of the sear block group.

(4) TOOL, RETAINING, SEAR BUFFER SPRING. This is a rod bent to form an oval handle at one end, slightly tapered at the end of the straight portion. The tapered end is inserted into the hole of the sear block to engage the grooves on the sear buffer spring plungers, and thus hold the

20-MM AUTOMATIC GUN M1 AND
20-MM AIRCRAFT AUTOMATIC GUN AN-M2

sear buffer springs under compression. The sear block group can then be removed or replaced in the receiver as a unit.

(5) TOOL, REMOVING, TUBE LOCKING PIN. This tool consists of a cylinder open at one end with a small threaded hole in the other end. A threaded rod with a handle is screwed into the threaded hole in the cylinder. The threaded rod is screwed into the tube locking pin to remove it.

(6) TOOL, UNLOCKING, BREECHBLOCK. This lever with two arms pivoted at one end is designed to place it on top of the bolt body in the receiver. The hook on one arm engages the front face of the right breechblock slide so that when the lever is operated, the slides are forced rearward to unlock the breechblock.

(7) WRENCH, ENGRS., SGLE.-HD. This open-end wrench is provided to fit the gas cylinder guide and gas cylinder vent plug.

(8) WRENCH, REAR BUFFER. One face of this tool has a hexagon socket to fit the driving spring guide head; the other face has four equally spaced projections to engage keyways in the flange of the rear buffer sleeve. The socket has transverse circular slots for accommodating the handle.

(9) WRENCH, SPANNER. This spanner wrench is used to turn the mounting sleeve nut in adjusting the compression of the recoil spring.

(10) WRENCH, MUZZLE BRAKE. This wrench has internal splines to engage the external splines on the muzzle brake when disassembling and assembling. The tool can also be used for removing and replacing the muzzle thread protector.

Section X

STORAGE AND SHIPMENT

	Paragraph
Preparation for storage and shipment	75
Packaging	76
Removal of preservatives	77

75. PREPARATION FOR STORAGE AND SHIPMENT.

a. Preparation of Parts. Prepare the gun for storage and shipment by inverting the inner bowden connection bushing so that this inner bushing fits into the outer bowden connection bushing. This is to prevent the bowden connection shaft from being operated and to prevent the bushing from being damaged. Remove the muzzle brake and place the thread protector in position.

b. Cleaning. Clean the gun with SOLVENT, dry-cleaning, or with soapy water so as to remove shop coating, dirt, and other foreign substances from all surfaces. Thoroughly dry the gun before application of COMPOUND, rust preventive, light.

c. Application of Rust Preventive Compound. Immediately after the gun is cleaned, brush or slush on the outside of the gun light COMPOUND, rust preventive, light.

76. PACKAGING.

a. Pack the gun in a box similar to that shown in fig. 69. Make this box of 1-inch lumber to the following dimensions:
Length ... 95 in.
Width ... 7¾ in.
Depth ... 8⅜ in.
The weight of the box with contents will be 158 pounds. The box shall be lined with waterproof paper and shall be strapped with either 3 round wire straps (No. 14 gage) or 3 flat steel straps (⅜ x 0.020 in.).

77. REMOVAL OF PRESERVATIVES.

a. Remove preservatives by cleaning all surfaces with SOLVENT, dry-cleaning.

TM 9-227
78-80

**20-MM AUTOMATIC GUN M1 AND
20-MM AIRCRAFT AUTOMATIC GUN AN-M2**

Section XI

OPERATION UNDER UNUSUAL CONDITIONS

	Paragraph
General	78
Care in arctic climates	79
Care in tropical climates	80

78. GENERAL.

a. When operating the gun under unusual conditions such as tropical or arctic climates, severe dust or sand conditions, and near salt water, the precautions listed below should be scrupulously followed.

79. CARE IN ARCTIC CLIMATES.

a. In arctic climates, it is essential that all moving parts be kept absolutely free of moisture. Clean and lubricate all parts but do not use excess lubricant because it may solidify to such an extent as to cause sluggish movement or even complete failure. When the gun is in the open, cover all unprotected parts with tarpaulin or other suitable material. Select a firm covering so that no loose material will get into the working parts of the gun. When the gun is transferred from the outside into a heated building, clean and oil it immediately to prevent the condensation of moisture. After the gun has reached room temperature, wipe it dry with a clean rag and oil again.

80. CARE IN TROPICAL CLIMATES.

a. In hot and tropical climates where humidities are high or where salt air is present, inspect and clean the gun frequently as required rather than at fixed intervals. Clean and oil as soon as possible after firing, when the gun gets wet or dirty, or if there is any reason to expect corrosion to start. In hot but dry climates where sand or dust are prevalent, clean the guns of all lubricant and leave entirely dry. Cover guns with a tarpaulin or other suitable protection. Lubricate immediately upon returning guns to normal climate.

TM 9-227
80

OPERATION UNDER UNUSUAL CONDITIONS

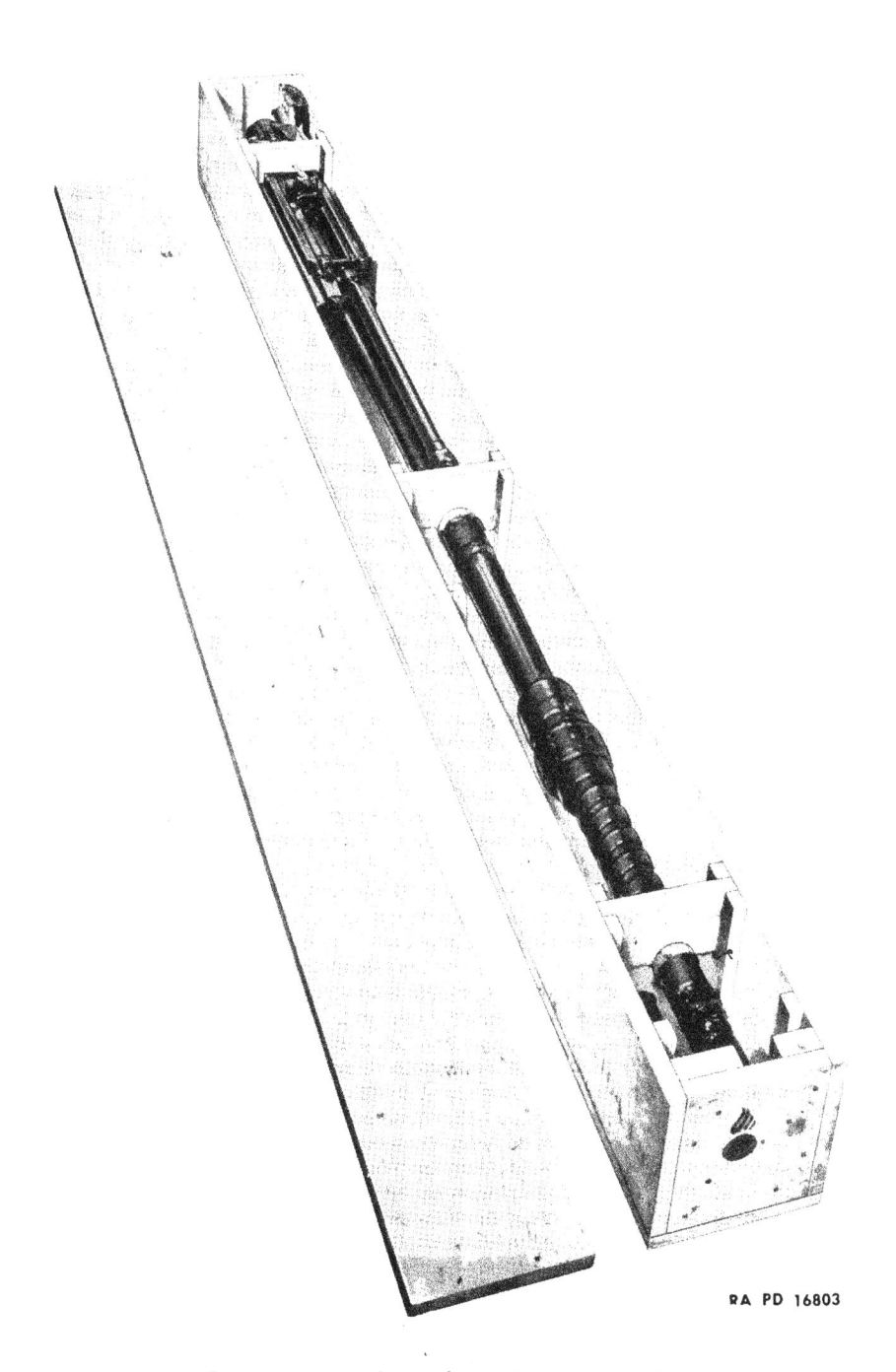

Figure 69 — Method of Packing 20-mm Gun

TM 9-227
81–82

20-MM AUTOMATIC GUN M1 AND
20-MM AIRCRAFT AUTOMATIC GUN AN-M2

Section XII

REFERENCES

	Paragraph
Standard nomenclature lists	81
Explanatory publications	82

81. STANDARD NOMENCLATURE LISTS.

   a. Ammunition, fixed and semifixed, all types for pack, light and medium field artillery ..... SNL R-1

   b. Cleaning, preserving and lubricating materials, recoil fluids, special oils, and similar items of issue ..... SNL K-1

   c. Gun, automatic, 20-mm, M1 and AN-M2 (aircraft) SNL A-47

   d. Soldering, brazing and welding material, gases and related items ..... SNL K-2

Current Standard Nomenclature Lists are as tabulated here. An up-to-date list of SNL's is maintained as the "Ordnance Publications for Supply Index" ..... OPSI

82. EXPLANATORY PUBLICATIONS.

   a. Air Corps Technical Order ..... T.O. 11-1-21

   b. Ammunition, general ..... TM 9-1900

   c. **Army Regulations.**
     Ordnance field service in time of peace ..... AR 45-30
     Range regulations for firing ammunition for training and target practice ..... AR 750-10

   d. Cleaning, preserving, lubricating, and welding materials and similar items issued by the Ordnance Department ..... TM 9-850

INDEX

A

	Page No.
Accessories (*See* Spare parts and accessories)	
Accidents, ammunition, field report of	90
Adapter	9–11
Adjustments (*See* Failures and corrections)	
Ammunition	
authorized rounds	83–86
list of	86
care, handling, and preservation	83
classification	79–80
field report of accidents	90
fuzes	88
general information on	79
identification	80
lot number	80
mark or model	80
marking	82
painting	80–82
lubrication of	6, 35
marking for shipment	90
nomenclature, standard	79
packing	88
preparation for firing	86
shell, fixed, H.E.I., Mk.1, w/fuze, percussion, etc.	86
shell, fixed, steel, Mk.1, etc.	87–88
shot, fixed, A.P., M75, w/tracer, etc.	87
Arctic climates, care of guns in	94
Assembly	
breechblock lock	60
driving spring and guide	60–65
feed mechanism, 20-mm, M1	
for left-hand feed	73
for right-hand feed	71–73
magazine, 60-round, M1	66
muzzle brake	65
sear block group	65
sear cover plate	65

B

Belt, feed mechanism, description and functioning	24
Blocks, inertia (*See* Inertia blocks)	
Bolt, description and functioning	15
Bore of tube, data on	4
Box, packing, dimensions and weight	93
Breechblock assembly	
description and functioning	13–15
disassembly	52

TM 9-227

	Page No.
functioning of in firing	25–34
inspection	75–76
removal	48–49
Breechblock lock	
assembly	60
description and functioning	15
Breechblock locking key	
description and functioning	15–19
disassembly	60
Buffer (*See* Rear buffer group)	
Bushing (mounting sleeve group)	
function	9
oiling	9

C

Care and preservation	
ammunition	83
cleaning instructions	
feed mechanism	44–45
gun(s)	44
preparing for storage and shipment	93
received from storage	45
lubrication instructions	45
replacement of parts	45
Cautions for using service	6
Cessation of firing	40
Characteristics of gun, auto., 20-mm, M1	3
Classification of ammunition	79–80
Cleaning instructions (*See under* Care and preservation)	
Cleaning staff M13 (20-mm)	92
Climates, arctic and tropical, care of guns in	94
Cocking the gun	38
Compound, rust preventive, application of after cleaning gun	93
Corrections after flight	42–43
(*See* Failures and corrections *for list*)	
Cycle, firing, description of	25–34

D

Data, tabulated	4
Description and functioning	8–34
breechblock assembly	13–15
breechblock locking key	15–19
driving spring guide group	21
feed mechanism, 20-mm, M1	24
functioning of the gun as a whole	25–34

TM 9-227

20-MM AUTOMATIC GUN M1 AND 20-MM AIRCRAFT AUTOMATIC GUN AN-M2

D—Cont'd   Page No.

Description and functioning—Cont'd
- gas cylinder and sleeve group.... 11
- magazine, 60-round, 20-mm, M1   25
- magazine slide group 13
- muzzle brake assembly......... 8–9
- rear buffer assembly 21
- receiver assembly............ 11–13
- recoil spring and mounting sleeve group 9–11
- sear block group 19
- sear cover plate group....... 19–21
- tube 8

Designations of guns 4
Differences between models of guns 3–4
Dimensions
- gun 4
- packing box 93

Disassembly
- breechblock group............ 52
- breechblock locking key...... 60
- driving spring guide group . 51–52
- feed mechanism, 20-mm, M1
  - for left-hand feed 73
  - for right-hand feed 66
- gas cylinder and sleeve group .. 60
- magazine, 60-round, M1 65–66
- magazine slide group 60
- muzzle brake assembly....... 60
- rear buffer group 52
- recoil spring and mounting sleeve group 60
- sear block group........... 56–59
- sear cover plate group...... 52–56

Driving spring guide group
- assembly 60–65
- description and functioning 21
- disassembly 51–52
- inspection 75

E

Eject, failure to 43
Ejector, description and functioning 13
Extract, failure to............... 43
Extractor spring, functioning of .. 29

F

Failures and corrections:
- of round to enter chamber in tube 43
- to eject 43
- to extract 43
- to feed................... 42–43
- to fire 43
- to unlock.................... 43

Page No.

Feed mechanism, 20-mm M1
- assembly
  - for left-hand feed. 73
  - for right-hand feed........ 71–73
- checking tension of driving spring 42
- cleaning instructions 44–45
- description and functioning..... 24
- disassembly (for right-hand feed) 66
- inspection 74
- loading 38–39
- belt
  - joining belts.............. 36
  - left-hand feed 36
  - right-hand feed 35–36
- unloading 41
- weight 4

Field report of ammunition accidents 90
Filler sleeve, function of. 9
Firing
- ammunition, preparation 86
- cycle, description.......... 25–34
- gun 40
- stopping 34
- inspection (See Inspection)
- method of in aircraft installations 21
- rate of...................... 4

Firing pin, malfunction and correction 43
Flight
- corrections to gun after....... 42–43
- immediate action to remedy stoppages in prohibited 42

Functioning of the gun as a whole 25–34
(See also Description and functioning)

Fuzes 88

G

Gas cylinder (and sleeve group)
- description 11
- disassembly 60
- functioning during firing ... 29
- inspection 78

I

Identification of ammunition 80
Immediate action to remedy stoppage in flight prohibited 42
Inertia blocks
- description 15
- function during firing 29

TM 9-227

INDEX

I—Cont'd

	Page No.
Inspection	74–78
breechblock group	75–76
driving spring guide group	75
feeder	74–75
gas cylinder and sleeve group	78
general	74
magazine slide group	77
muzzle brake group	78
rear buffer group	75
receiver group	9
recoil spring and mounting sleeve group	78
sear block group	76–77
sear cover plate group	76
tube	78

J

Joining new belt to partly expended belt ... 36

K

Key, breechblock locking (See Breechblock locking key)

L

Last round, removal of from 20-mm feed mechanism	41
Latch, magazine, operation	13
Loading	
belt for feed mechanism, 20-mm, M1	
joining new belt to partly expended belt	36
left-hand feed	36
right-hand feed	35–36
gun (See gun under Operation)	
Lock, breechblock, description and functioning	15
Locking key, breechblock (See Breechblock locking key)	
Lot number of ammunition	80
Lubricants, care of	44
Lubrication instructions	45
ammunition	6, 35

M

Magazine, 60-round, M1	
assembly	66
cleaning instructions	44–45
description and operation	25
disassembly	65–66
inspection	74–75
loading	40
applying initial tension	36–37

	Page No.
rounds, checking position of	42
unloading	40–41
weight	4
Magazine slide group	
description and functioning	13
disassembly	60
inspection	77
Malfunction of ammunition	90
(See also Failures and corrections)	
Mark or model of ammunition	80
Marking ammunition	82
for shipment	90
Models of gun	3–4
Mounting of gun	3
Mounting sleeve assembly	
description and functioning of	9
disassembly	60
inspection	78
Muzzle brake assembly	
description and functioning	8–9
disassembly	60
inspection	78
replacement	65
removal	51
Muzzle velocity	4

N

Nomenclature, standard, use of for ammunition ... 79

O

Oiling	
bushing of mounting sleeve group	9
Operation	35–41
ammunition, lubrication of	35
feed mechanism, 20-mm, M1, loading a belt for	35–36
joining belts	36
left-hand feed	36
right-hand feed	35–36
gun	
cocking	38
firing	40
loading	
feed mechanism, 20-mm, M1	38–39
magazine, 60-round, M1	40
applying initial tension	36–37
unloading	
feed mechanism, 20-mm, M1	41
magazine, 60-round, M1	40
operation under unusual conditions	94

99

TM 9-227

20-MM AUTOMATIC GUN M1 AND
20-MM AIRCRAFT AUTOMATIC GUN AN-M2

O—Cont'd Page No.

Organization spare parts and accessories (See Spare parts and accessories)
Over-all length of gun . 4

P

Packaging for storage and shipment 93
Packing ammunition 88
Painting ammunition 80–82
Parts
  preparation of for storage and shipment 93
  replacement of 45
Piston, dashpot, function of in recoil 9
Plungers, inertia block, function of 15
Preservation (See Care and preservation)
Preservatives, removal of 93
Projectile
  movement of in firing 25
  travel of in tube 4

R

Rate of fire 4
Rear buffer group
  description and function 21
  disassembly 52
  inspection 75
  removal 48
Receiver assembly
  description and functioning 11–13
  differences in parts 4
  inspection 78
Recoil
  adjusting compression of recoil spring 42
  breechblock action during 29
  description 9–11
  to operate M1 feed mechanism 4, 24
Recoil mechanisms, types 9–11
Recoil spring
  adjusting compression of 42
  inspection 78
Recoil spring and mounting sleeve group
  description and functioning 9–11
  disassembly 60
Retainer assembly, description and functioning 21

Page No.

Removal of groups and assemblies 46–51
  breechblock group 48–49
  driving spring guide group 46–48
  muzzle brake group 51
  rear buffer group 48
  sear block group 49–51
  sear cover plate group 49
Removal of last round from 20-mm feed mechanism 41
Replacement (See Assembly)
Replacement of parts 45
Rifling, data on 4
Rounds
  authorized 83–86
  table 86
  checking of in magazine 42
  failure of to enter chamber 43
  lubrication before inserting in magazine or belt 6
Run-away gun, correction 43
Rust preventive compound, application of after cleaning gun 93

S

Sear
  action in automatic fire 29
  action in stopping firing 34
Sear block group
  assembly 65
  description and functioning 19
  disassembly 56–59
  inspection 76–77
  removal 49–51
Sear cover plate group
  assembly 65
  description and functioning 19–21
  disassembly 52–56
  inspection 76
  removal 49
Shell, fixed, H.E.I., Mk. 1, w/fuze, percussion 86
Shell, fixed, steel, Mk. 1 87–88
Shipment (See Storage and shipment)
Shipment, marking ammunition for 90
Shot, fixed, A.P., M75, w/tracer 87
Sleeve mounting group
  description 11
  disassembly 60
  inspection 78
Slides, breechblock, description and functioning 15

100

INDEX

S—Cont'd	Page No.
Solenoid, firing with in aircraft installations	21
Spare parts and accessories	91–92
accessories	
general (gun)	91
staff, cleaning, M13 (20-mm)	92
tool, assembling, driving spring	91
tool, assembling, sear block	92
tool, breechblock, unlocking	91
tool, removing, tube locking pin	92
tool, retaining, sear buffer spring	91–92
wrench, engrs., sgle.-head	92
wrench, muzzle brake	92
wrench, rear buffer	91
wrench, spanner	92
organization spare parts	91
Spring dimensions, table of	77
Spring, driving, replacement	45
Staff, cleaning, M13 (20-mm)	92
Standard nomenclature, use of for ammunition	79
Stoppages	
caution in remedying	6
immediate action to remedy prohibited in flight	42
(See Failures and corrections for applicable actions)	
Stopping firing of gun	34, 40
Storage and shipment	93
application of rust preventive compound	93
cleaning	93
guns received from storage	45
packaging	93
box, dimensions and weight	93
preparation of parts	93
removal of preservatives	93

	Page No.
Tables	
authorized rounds of ammunition	86
spring dimensions	77
Tabulated data on gun	4
Tools	
breechblock, unlocking	91
driving spring, assembling	91
sear block, assembling	92
sear buffer spring, retaining	91
special	46
tube locking pin, removing	92
Trigger mechanism, description of	19, 21
Tropical climates, care of guns in	94
Tube	
bore, data on	4
description	8
inspection	78
weight and length	4

U

Unloading	
feed mechanism, 20-mm, M1	41
magazine, 60-round, M1	40–41
Unlock, failure to	43

V

Velocity, muzzle	7

W

Weight	
gun	4
packing box for	93
Wrenches	
engrs., sgle.-hd.	92
muzzle brake	92
rear buffer	91
spanner	92

```
A.G. 062.11 (6-29-42)
O.O. 461/14553 O.O. (11-9-42)
```

BY ORDER OF THE SECRETARY OF WAR:

G. C. MARSHALL,
Chief of Staff.

OFFICIAL:
    J. A. ULIO,
        Major General,
        The Adjutant General.

Distribution: D 1(2); IB 1(2); IR 1(2); IBn 1(5); 9(2); IC 1(2); 9(4).
(For explanation of symbols, see FM 21-6)

Lightning Source UK Ltd.
Milton Keynes UK
UKHW02f2016040618
323724UK00041B/1584/P